Mark Anthony Benvenuto
Industrial Inorganic Chemistry

Also of Interest

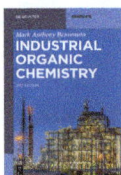

Industrial Organic Chemistry
Benvenuto, 2024
ISBN 978-3-11-132991-8, e-ISBN (PDF) 978-3-11-133035-8,
e-ISBN (EPUB) 978-3-11-133080-8

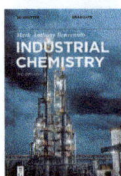

Industrial Chemistry
Benvenuto, 2023
ISBN 978-3-11-067106-3, e-ISBN (PDF) 978-3-11-067109-4,
e-ISBN (EPUB) 978-3-11-067121-6

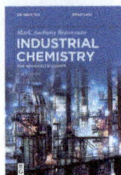

Industrial Chemistry
For Advanced Students
Benvenuto, 2023
ISBN 978-3-11-077874-8, e-ISBN (PDF) 978-3-11-077876-2,
e-ISBN (EPUB) 978-3-11-077898-4

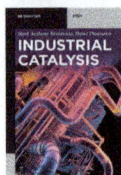

Industrial Catalysis
Benvenuto, Plaumann, 2021
ISBN 978-3-11-054284-4, e-ISBN (PDF) 978-3-11-054286-8,
e-ISBN (EPUB) 978-3-11-054294-3

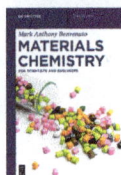

Materials Chemistry
For Scientists and Engineers
Benvenuto, 2022
ISBN 978-3-11-065673-2, e-ISBN (PDF) 978-3-11-065677-0,
e-ISBN (EPUB) 978-3-11-065682-4

Mark Anthony Benvenuto

Industrial Inorganic Chemistry

2nd Edition

DE GRUYTER

Author
Prof. Dr. Mark Anthony Benvenuto
Department of Chemistry and Biochemistry
University of Detroit Mercy
4001 W. McNichols Road
Detroit, MI 48221-3038
USA

ISBN 978-3-11-132944-4
e-ISBN (PDF) 978-3-11-132951-2
e-ISBN (EPUB) 978-3-11-132963-5

Library of Congress Control Number: 2024931629

Bibliographic information published by the Deutsche Nationalbibliothek
The Deutsche Nationalbibliothek lists this publication in the Deutsche Nationalbibliografie;
detailed bibliographic data are available on the Internet at http://dnb.dnb.de.

© 2024 Walter de Gruyter GmbH, Berlin/Boston
Cover image: maki_shmaki/iStock/Getty Images Plus
Typesetting: Integra Software Services Pvt. Ltd.
Printing and binding: CPI books GmbH, Leck

www.degruyter.com

Preface

The large-scale production of several commodity chemicals, fertilizers, and metals has changed the modern world in ways not imagined throughout most of history. Human life spans and quality of life have been extended and improved radically because of our ability to mass produce chemicals such as sulfuric acid and calcium carbonate, ammonia-based fertilizers, as well as refine and alloy numerous different metals. These developments have given us a steady food supply, reproducibly mass produced medicines, ready supplies of clean water, ease and speed of communications and transportation, and an enormous variety of products and services that were unimaginable ever before in history. This book tries to survey the production of many of the elements and compounds which make such advances possible, and stimulate thought on how this can be made sustainable and as environmentally friendly as possible.

Writing this book has been educational, challenging, and rewarding. An endeavor of this sort is never really done alone, so I would like to thank my editors, Karin Sora, Julia Lauterbach, Kathleen Prüfer, and Ria Fritz for all their help, advice, and encouragement. I also wish to thank several of my work colleagues and friends – Matt Mio, Liz Roberts-Kirchhoff, Kate Lanigan, Klaus Friedrich, Kendra Evans, Schula Schlick, Jon Stevens, Mary Lou Caspers, Bob Ross, Prasad Venugopal, Gary Hillebrand, Jane Schely, and Meghann Murray, all of whom endured my questions about one subject or another without even realizing I was gauging their answers for some aspect of what I was writing. As well, thanks go to colleagues and friends at BASF, especially Heinz Plaumann and Denise Grimsley for letting me pick their brains on several subjects. And a very special thank you goes to Megan Klein of Ash Stevens for her help, and for proofreading these chapters as I wrote them.

Finally, as always, I must thank my wife Marye, and my sons David and Christian, for putting up with all my strange queries, odd hours and late night writing sessions.

Detroit, September 2015
Mark A. Benvenuto

https://doi.org/10.1515/9783111329512-202

Contents

1 Overview and introduction to industrial inorganic processes

The common lore in the 21st century is that in Ancient Greece, the philosophers of the day believed that all materials in the world were made from the four elements: earth, water, air, and fire. Materials such as wood were explained as being a mixture of some amount of probably earth, water, air, and possibly even fire (that had not yet been released). Perhaps ironically, even in the chemically complex world in which we live today, all our materials can be traced back to sources that come from the earth, the water, and the air – and many of them are transformed with fire of some sort.

It is always difficult to delineate the sources of what get called the major chemicals because it is difficult to determine what constitutes "major", and because there are sometimes multiple sources for the same material. For example, the amount of iron produced annually on a national and a global scale is tracked by several organizations such as the United Nations and the United States Geological Survey (USGS), and is usually recorded in thousands of metric tons [1, 2]. Another metal, gold, is also tracked by organizations including the USGS and the World Gold Council, but is measured in tons, and is priced in ounces [2,3]. As far as materials that are derived from different sources, sulfur can be extracted from in-ground deposits through what is called the Frasch process, but it is also recovered from oil-refining operations. In both cases, the sulfur is used for the same end product – sulfuric acid [4].

Materials that are mined

Numerous materials that are used in some chemical process or another, or that ultimately are formed into some end-user product, are mined. The term mining often implies certain processes, such as the removal of a hilltop and creation of a large pit, or digging a deep shaft into the earth to extract some metal or ore. But mining can also include inserting pipes into the ground and using hot solutions or pressurized liquids or gases to extract a material from the ground. This book contains examples of materials that are obtained through all of these methods.

Materials from water

Numerous reactions must be run in water, but in several other cases, large-scale chemistry is performed that uses water as one of the reactants. The production of sulfuric acid, as well as of three other large commodity chemicals: sodium hydroxide, elemental chlorine, and elemental hydrogen – known as the chlor-alkali process – are examples of such processes.

https://doi.org/10.1515/9783111329512-001

Inorganics extracted from organic sources

Perhaps the most difficult processes to categorize neatly are those in which some inorganic material is produced from an organic one, or in which some inorganic product depends upon an organic one for its production. The large-scale production of sulfuric acid can have an organic source of sulfur. The large-scale production of carbon black represents another material that is generally defined as inorganic that requires an organic feedstock.

Materials from air

Even many chemists and chemical engineers do not often think of air as a starting material for chemical transformations and chemical production. Yet air provides oxygen and nitrogen, as well as carbon dioxide and argon, all of which can be involved in further chemical reactions. Air liquefaction plants provide vital starting materials for processes that make sulfuric acid, ammonia, and nitric acid, to name just a few of the larger processes.

List of producers by country

The USGS claims in their annual Mineral Commodity Survey that the economic health of a nation can be measured by its production of sulfuric acid [2]. In this book, we mention and examine the geographic sources for all the materials in the different chapters. While some materials are wide spread across the planet, others are much more localized. These localized source materials are never used to determine the economic health of a nation. But an economically weak nation cannot generally afford to extract, refine, and produce such commodities.

This book discusses the major inorganic chemicals that are used in industry, and also tries to discuss the possibilities for recycling and re-use of these materials. In every case, time, energy, and money are required to produce these commodity chemicals and materials. This is because it is often more economically sound to re-use or in some way recycle a material when the item in which it is used reaches the end of its usable life span [5].

Bibliography

[1] United Nations, UN ComTrade. Website. (Accessed 18 December 2023, as: https://comtradeplus.un.org).
[2] United States Geological Survey. Website. (Accessed 18 December 2023, as: https://pubs.usgs.gov).
[3] World Gold Council. Website. (Accessed 18 December 2023, as: https://www.gold.org).
[4] Sulfuric Acid Today. Website. (Accessed 18 December 2023, as: https://h2so4today.com).
[5] United States Environmental Protection Agency. Website. (Accessed 17 November 2014, as: https://www.epa.gov).

2 Sulfuric acid production, uses, derivatives

2.1 Introduction

The production of sulfuric acid does not come readily to mind when a person thinks of a chemical or material that they use on a daily basis. Yet this material has many uses, either in the production of other bulk chemicals, or ultimately in the production of some user end products.

The production of sulfuric acid has been linked to the economic health of a developed nation. The United States Geological Survey (USGS) annual Mineral Commodity Summaries [1] does not specifically track sulfuric acid, only because it must be made from another material, namely sulfur. The Mineral Commodity Summaries 2013 does track sulfur production, and comments that in the recent past, "recovered elemental sulfur and byproduct sulfuric acid were produced at 95 operations in 27 States" [1]. Clearly, the production of sulfur and sulfuric acid is a large, widespread operation. Such a statement also implies that sulfuric acid production is the major use of elemental sulfur.

2.2 Sulfur sourcing

For the last century, sulfur has been mined from underground deposits via what is called the Frasch process. This involves inserting three concentric tubes into the ground and into the sulfur deposit, blowing superheated water into the deposit through the outermost tube, blowing hot air into the central tube, and thus forcing out the water–sulfur mixture. The air needs to be blown into the mix because the sulfur–water mixture is denser than water, and it will not rise without this increased pressure. This is shown in Figure 2.1.

Sulfur is also obtained as a by-product of metal refining from sulfide ores. The roasting of ores had, in the past, released large amounts of sulfur oxides, but with increasing environmental awareness that these gases can be major sources of pollution, they have been captured and used.

In recent years, increasing amounts of sulfur are obtained in the form of hydrogen sulfide from refining the lightest fractions of crude oil. In what is called the Claus process, this is converted to sulfur, which is then used to produce sulfuric acid. Scheme 2.1 shows the reaction chemistry for the production of sulfur via this method in a simplified form.

Feed gases generally need at least 25 % H_2S for this recovery to be economically feasible. Also, the reaction must be run at approximately 850 °C, which means that the cost of the energy involved must be factored into determining economic viability. This is also the means by which the majority of sulfur is now obtained for further use in the production of sulfuric acid. While this represents an organic source of the element sulfur,

https://doi.org/10.1515/9783111329512-002

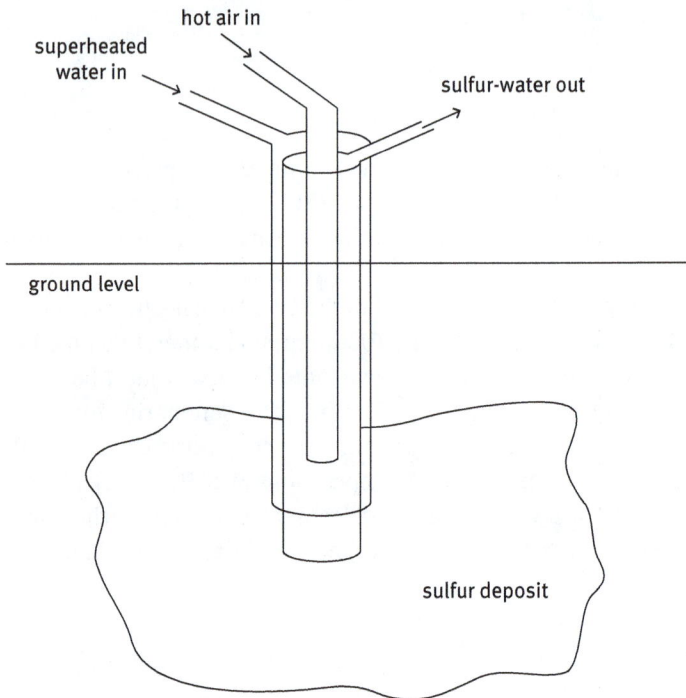

Figure 2.1: Frasch process.

$$O_{2(g)} + 2\,H_2S_{(g)} \longrightarrow 2\,S + 2\,H_2O$$

Scheme 2.1: Sulfur recovery from natural gas.

its subsequent use in the production of sulfuric acid and any other sulfur-containing compounds is generally considered to be inorganic process chemistry.

2.3 Sulfuric acid, methods of production

When produced from elemental sulfur, the reaction chemistry for the production of sulfuric acid can be broken down into five steps, as shown in Scheme 2.2. This is called the contact process.

The second step is catalyzed with vanadium pentoxide (V_2O_5), a catalyst that often has a working lifetime of up to 20 years. The reaction runs between 400 °C and 600 °C, and the catalyst is often activated by the addition of potassium oxide. The reaction is not stopped at the first production of sulfuric acid, the third reaction in Scheme 2.2, because the direct addition of sulfur trioxide to water produces a corrosive mist. Rather, sulfur

$$S + O_{2(g)} \longrightarrow SO_{2(g)}$$
$$SO_{2(g)} + \tfrac{1}{2}\,O_{2(g)} \longrightarrow SO_{3(g)}$$
$$SO_{3(g)} + H_2O \longrightarrow H_2SO_4$$
$$H_2SO_4 + SO_{3(g)} \longrightarrow H_2S_2O_7$$
$$H_2S_2O_7 + H_2O \longrightarrow 2\,H_2SO_4$$

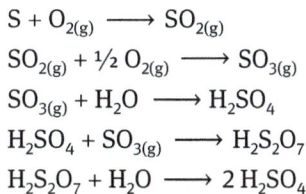

Scheme 2.2: Sulfuric acid production.

trioxide is absorbed into existing aqueous concentrated sulfuric acid, forming what is still called oleum ($H_2S_2O_7$). The final reaction is the addition of water to this, to form concentrated sulfuric acid.

2.4 Sulfuric acid, annual volume of production

Roughly 309.24 million tons… In 2023, production was up significantly from 2022, largely because of the removal of COVID restrictions in populations, and the resumption of work on a large scale [2, 3].

2.5 Sulfuric acid uses

Overwhelmingly, sulfuric acid is used to produce fertilizers. Phosphate fertilizer production is intimately tied to the use of sulfuric acid through a very mature, large-scale process.

Scheme 2.3 shows the simplified reaction chemistry whereby phosphoric acid, as well as hydrofluoric acid, is made from sulfuric acid.

$$Ca_5F(PO_4)_3 + 5\,H_2SO_4 + 10\,H_2O \longrightarrow 5\,CaSO_4 \cdot 2\,H_2O + 3\,H_3PO_4 + HF$$

Scheme 2.3: Phosphoric acid production.

Other uses include the production of numerous sulfur-containing chemicals – some of the most common of which are discussed in Section 2.6 – that are further used in some chemical transformation. Metal processing and petroleum refining are two important uses for sulfuric acid. One consumer end use of sulfuric acid that is fairly well known is that of lead acid batteries, where a sulfuric acid solution is required for the redox chemistry of the battery to function.

2.6 Derivatives

2.6.1 Sulfur dioxide

Sulfur dioxide is usually not isolated, most of it being consumed in the production of sulfuric acid. However, there are still some uses for sulfur dioxide itself. It is used as a food preservative, specifically a fruit preservative, and assigned the number E220 as a European Union food additive. At least one web site devoted to food and food additives, "Better Health Channel", points out that sulfur dioxide has been prohibited from use on fruits and vegetables in the United States, citing links between its use and bronchial disorders, especially in people who suffer from asthma [4].

Sulfur dioxide, at the level of parts per million, can serve as an anti-oxidant and anti-microbial material in various wines. The words, "contains sulfites" is often written on the ingredients lists on bottles of wines, in the event that a consumer is sensitive to them.

Traditionally, sulfur dioxide was also used as a refrigerant, but the material was replaced by chlorofluorocarbons for use in personal refrigerators.

2.6.2 Sulfur trioxide

Almost all sulfur trioxide is used in the production of sulfuric acid, as shown in Figure 2.1. A much smaller amount is used in cleaning flue gases, because sulfur trioxide mixed with particulate matter imparts a charge to the particles. This then results in particulate material that can be trapped by electrostatic precipitators, and not emitted to the environment.

2.6.3 Hydrogen sulfide gas

This simple compound is known to many people as the "rotten egg gas", because of its foul odor, which can be detected at very low concentrations. While it can be produced from the elements, hydrogen sulfide can also be extracted from crude oil, as mentioned.

While some hydrogen sulfide is added to the gas used in heating homes, so that a gas leak is easily detectable, much more is used to remove metal ions from aqueous solutions. This can be applied to potable waters, but is often used to remove metals from water during froth floatation extraction techniques, because the H_2S converts metals into metal sulfides, which usually have low solubilities.

2.6.4 Sodium sulfide

Sodium sulfide is another sulfur-containing compound that finds uses that can be classified as inorganic or organic. Its production can be written in a straightforward manner, as shown in Scheme 2.4.

$$2\,C + Na_2SO_4 \longrightarrow Na_2S + 2\,CO_2$$

Scheme 2.4: Sodium sulfide production.

Many of the sources of carbon are considered organic. The direct reaction of the two elements will also produce sodium sulfide, but is not economically as feasible as the reduction of sodium sulfate.

Most sodium sulfide finds use in the Kraft process for paper production. Kraft is the German word meaning strong, and is unrelated to the food manufacturing company. Wood pulp must have the cellulose and lignin separated so that paper can be made from it. A mixture of NaOH, NaSH, and Na_2S, called white liquor, is used in the pressure digestion step of the process [5, 6]. The wood chips and this mixture are cooked for 2–5 h at 7–9 atm and 175 °C. The pulp is then separated from this aqueous mixture.

Like hydrogen sulfide, sodium sulfide can also be used to remove metal ions from aqueous solutions by forming insoluble metal sulfides.

2.6.5 Carbon disulfide

This colorless liquid is produced on a roughly million-ton scale annually. The general reaction chemistry that illustrates this is as follows in Scheme 2.5.

$$3S + CH_4 \longrightarrow CS_2 + 2\,H_2S$$

Scheme 2.5: Carbon disulfide production.

While this reaction appears to be straightforward, it must be run at approximately 600 °C, and must utilize an alumina or silica catalyst.

The applications of carbon disulfide are wide, and include its use as a solvent. Broadly, they can be categorized as follows:
1. Insecticide – carbon disulfide has proven to be effective as an insecticide in grain storage and as a soil insecticide.
2. Fumigant – carbon disulfide has been found to be an effective fumigation material in airtight environments such as long-term storage warehouses.

3. Chemical production – carbon disulfide is used in the manufacture of carbon tetrachloride and polymers such as rayon.
4. Solvent – there are a variety of reactions that have been found to run successfully in carbon disulfide. Often these reactions involve a sulfur-containing reactant.

There are other uses for carbon disulfide as well.

2.6.6 Sulfur chlorides

The two sulfur chlorides, sulfur dichloride (SCl_2) and disulfur dichloride (S_2Cl_2), are made by the direct chlorination of sulfur; with SCl_2 being produced through the addition of further chlorine to S_2Cl_2. Both have been used to prepare what is called "sulfur mustard", a poison gas that some militaries have stockpiled as a chemical weapon. The reaction chemistry, sometimes called the Levinstein process or the Depretz method depending on whether S_2Cl_2 or SCl_2 is used as a starting material, can be summed up, as shown in Scheme 2.6, in two reactions.

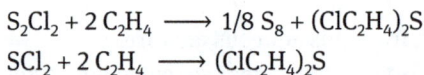

$$S_2Cl_2 + 2\,C_2H_4 \longrightarrow 1/8\,S_8 + (ClC_2H_4)_2S$$
$$SCl_2 + 2\,C_2H_4 \longrightarrow (ClC_2H_4)_2S$$

Scheme 2.6: The Levinstein process for sulfur mustard production.

Disulfur dichloride is also reacted with various aniline derivatives to produce thioindigo dyes, or their precursor molecules.

2.6.7 Thionyl chloride

Thionyl chloride, $SOCl_2$, is another inorganic, sulfur compound that has a large array of applications that are essentially organic. The synthesis of it can be represented fairly simply, as shown in Scheme 2.7.

$$SO_3 + SCl_2 \longrightarrow SOCl_2 + SO_2$$

Scheme 2.7: Thionyl chloride production.

The reaction represents the large-scale production of thionyl chloride. There are several other methods that can be used, but the above reaction is the major route.

As mentioned, thionyl chloride is used in numerous types of organic reactions. One class of inorganic reactions in which thionyl chloride has proven useful is the dehydration of metal chloride hydrates. The general reaction, using M for a metal ion, is seen in Scheme 2.8.

$$3\ SOCl_2 + MCl_3 \cdot 3\ H_2O \longrightarrow MCl_3 + 6\ HCl + 3\ SO_2$$

Scheme 2.8: Metal chloride dehydration with thionyl chloride.

Perhaps obviously, the by-products of such reactions need to be captured.

Thionyl chloride also serves as a cathode in some lithium rechargeable batteries. Lithium batteries have gained interest in recent years because of their high charge density, operational ability over a range of temperatures, and long shelf lives.

2.6.8 Sulfuryl chloride

This sulfur-containing compound finds its largest volume use in the production of pesticides. Because it is a liquid, it is often easier to use than chlorine gas. Sulfuryl chloride is often used as a means of delivering chlorine into some reaction.

The synthesis of sulfuryl chloride can be shown simply in Scheme 2.9. The reaction requires a catalyst. Activated carbon is often used for this purpose.

$$Cl_2 + SO_2 \longrightarrow SO_2Cl_2$$

Scheme 2.9: Sulfuryl chloride production.

Much like thionyl chloride, sulfuryl chloride also has a wide range of uses as a reagent in organic chemical reactions.

2.6.9 Chlorosulfonic acid (or, chlorosulfuric acid)

The major production route for chlorosulfonic acid, HSO_3Cl, involves reacting hydrochloric acid and sulfur trioxide, as shown in Scheme 2.10.

$$SO_3 + HCl \longrightarrow HSO_3Cl$$

Scheme 2.10: Chlorosulfonic acid production.

The major use of the material is in the production of detergents, which involves the reaction of the acid with some alcohol. Scheme 2.11 shows the simplified reaction chemistry for this.

$$HSO_3Cl + ROH \longrightarrow HSO_3\text{-}O\text{-}R + HCl$$

Scheme 2.11: Detergent production with chlorosulfonic acid.

This reaction produces a wide variety of molecules that possess the same polar, hydrophilic group, the HSO_3-head unit, and a variety of nonpolar, hydrophobic tail groups.

2.6.10 Sodium thiosulfate

Differing from sodium sulfate only by the replacement of a sulfur atom for an oxygen atom, sodium thiosulfate ($Na_2S_2O_3$) is often produced from sodium sulfide.

Several uses exist for sodium thiosulfate, including the following:

a. Gold recovery
In extracting gold from materials that contain only small amounts of it, gold cations are complexed using the thiosulfate anion. While its use in forming gold complexes is environmentally friendlier than the use of cyanide compounds, the $[Au(S_2O_3)_2]^{3-}$ complex is not as readily recovered with activated carbon.

b. Photography
Sodium thiosulfate is one of the well-established fixers in photographic development. This use has declined in recent years as digital photography coupled with computer printing has made inroads into personal and professional photography.

c. Medical treatment
Sodium thiosulfate has been found to be an effective treatment for cyanide poisoning. It has also been used as a disease-specific antifungal agent.

d. Water treatment
Sodium thiosulfate finds use as a water de-chlorinating agent. This application is used when treated waste water is released into local waterways.

2.6.11 Ammonium thiosulfate

Ammomium thiosulfate differs from the above-mentioned sodium thiosulfate only in the cation. The reaction chemistry for its production is shown in Scheme 2.12.

$$(NH_4)_2SO_3 + S \longrightarrow (NH_4)_2S_2O_3$$

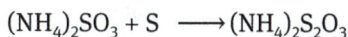

Scheme 2.12: Ammonium thiosulfate production.

The reaction also requires a nonstoichiometric amount of sulfide ions for it to run smoothly to completion, usually ammonium sulfide. It runs in a wide temperature range, can be run at lower temperatures, and tends to proceed to completion when an excess of ammonium hydroxide is also present.

The uses of ammonium thiosulfate and sodium thiosulfate are similar in regards to photography and gold reclamation. Ammonium thiosulfate can also be used for silver extraction, if necessary. It also finds use as a fertilizer, as it delivers both nitrogen and sulfur to the soil.

2.6.12 Sodium dithionite

This sodium–sulfur salt is prepared on a large scale according to the reaction shown in Scheme 2.13.

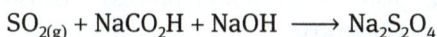

$$SO_{2(g)} + NaCO_2H + NaOH \longrightarrow Na_2S_2O_4$$

Scheme 2.13: Sodium dithionite production.

The process must be carried out at 65–80 °C, and requires the use of methanol as a solvent.

Sodium dithionite is used in various dyeing processes, including the dyeing of leather, wool, and cotton fabrics, because of its water solubility. It finds use in several other industries, the most prominent of which is arguably water treatment and purification. Its use in enhanced oil recovery has also grown significantly in recent years.

2.6.13 Sodium hydroxymethylsulfinate

This sodium salt, $NaSO_2CH_2OH$, was known by the trade name Rongalite™ when it was manufactured by BASF. Its manufacture requires formaldehyde, and the reaction chemistry can be simplified to that shown in Scheme 2.14.

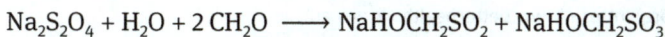

$$Na_2S_2O_4 + H_2O + 2\,CH_2O \longrightarrow NaHOCH_2SO_2 + NaHOCH_2SO_3$$

Scheme 2.14: Production of sodium hydroxymethylsulfinate.

The product is not stable for long periods. To produce a material that is stable and can be marketed, formaldehyde is added to the final product. Formaldehyde establishes the equilibrium of the reaction toward that of the sodium hydroxymethylsulfinate.

The $NaSO_2CH_2OH$ produced today is largely used in polymerizations, specifically in emulsion polymerizations. Originally, it was produced on an industrial scale for use as a bleaching agent.

2.6.14 Sodium hydrosulfide

Sodium hydrosulfide is used in the synthesis of several organic and inorganic compounds, usually in an aqueous environment. The simplified reaction chemistry for its production is shown in Scheme 2.15.

$$H_2S + NaOH \longrightarrow NaHS + H_2O$$

Scheme 2.15: Sodium hydrosulfide production.

Most NaHS is used in paper manufacturing, in the just-mentioned Kraft process. It is also used to help remove hair from hides in the leather industry.

2.7 Recycling

Sulfuric acid is produced so inexpensively that it is not recycled. Rather, much of the sulfuric acid that becomes a user end product is neutralized with base and discarded. Most of the sulfur-containing derivatives discussed in this chapter are made for use and distribution either into further chemical processes or into various user end products, and thus are not recycled. The Kraft process is run on a large enough scale that it is economically feasible to recover as much of the white liquor components (called 'black liquor' after pressure digestion) as possible.

Bibliography

[1] United States Geological Survey, Mineral Commodity Summaries, 2023. Website. (Accessed 18 December 2023 as: https://www.usgs.gov, as a downloadable pdf).
[2] Sulfuric Acid Today. Website. (Accessed 18 December 2023, as: https://h2so4today.com).
[3] Essential Chemical Industry. Website. (Accessed 18 December 2023, as: https://www.essentialchemicalindustry.org). https://www.betterhealth.vic.gov.au

[4] Better Health Channel. Website. (Accessed 18 December 2023, as: https://www.betterhealth.vic.gov.au). And Top 10 Food Additives to Avoid. Website. (Accessed 18 December 2023, as: https://www. hungryforchange.tv).

[5] American Forest & Paper Association. Website. (Accessed 18 December 2023, as: https://www. afandpa.org).

[6] Confederation of European Paper Industries, CEPI. Website. (Accessed 18 December 2023, https://www.pefc.org).

3 Industrial gases, isolation, and uses

3.1 Introduction

Both nitrogen and oxygen, the two major components of the air, are useful in several large-scale chemical transformations, and are useful as pure gases. These two, as well as argon, are all obtained from the liquefaction of air, which is a mature process, having been patented in 1895 [1]. The general steps for the liquefaction and separation of air always depend on the repeated expansion and contraction of the gaseous mixture, and the cooling that goes on at each step. Once air is liquefied, it can be separated into elemental and other gases based on the boiling points of each of the components. Table 3.1 lists the boiling points and relative abundance of the components of air.

Table 3.1: Components of air.

Component	Boiling Point (K)	Abundance
Nitrogen	77.4	78 %
Oxygen	90.2	21 %
Argon	87.3	0.93 %, 9,300 ppm
Carbon dioxide	194.7*	0.04–0.05 %, 400–500 ppm
Neon	27.1	18 ppm
Krypton	119.9	1 ppm
Xenon	165.1	87 ppb

*CO_2 sublimes at 1 atm.

The two processes that separate air into its components are the Linde process and the Claude process. Both take advantage of the fact that when air is compressed, it can be expanded into a larger volume, and then it cools. Ultimately, an oxygen-rich liquid is formed, which is further separated upon heating to the boiling point of nitrogen, so that it can be boiled and then re-condensed. Both of these are engineering processes that function at cryogenic temperatures, for which it is difficult to draw reaction chemistry.

3.2 Uses

3.2.1 Oxygen

Elemental oxygen is used in large amounts in the refining of metals, specifically iron. Oxygen is blown into the molten iron to aid in capturing sulfur impurities, and thus in purifying the molten material.

https://doi.org/10.1515/9783111329512-003

The large uses of elemental oxygen can be listed as follows [2–5]:
1. Iron refining
2. Production of ethylene
3. Metal cutting and welding, in oxy-acetylene torches
4. Medical applications

There are several smaller applications as well, and the amounts used fluctuate somewhat from one year to the next, depending upon the outputs of the products that involve oxygen. Additionally, oxygen is used in many processes and reactions for which it is not considered the main product. We have seen this already in the production of sulfuric acid, have just mentioned it in the production of iron, and will encounter it in several later chapters of this book.

3.2.2 Nitrogen

Most elemental nitrogen is used in the production of ammonia and ammonia-based fertilizers. This is discussed in the next chapter. When the USGS Mineral Commodity Summaries tracks the production and use of what it calls nitrogen, what is meant is actually ammonia, and does indicate so [6].

Nitrogen is also used as an inert gas blanket in a variety of processes, from large to small. Examples of the use of elemental nitrogen include the following:
1. Food preservation
2. Steel manufacturing
3. Blanketing gas
4. Filling automotive and truck tires

While nitrogen has been found to possess some reactivity, with reactive metals such as lithium and magnesium, and with some transition metal complexes, none of these resulting materials has yet become industrially significant.

Since liquefied nitrogen (sometimes abbreviated LN_2) is made in large-scale quantities, liquid nitrogen also has several applications. This include:
1. Coolant – in numerous processes and environments
2. Dermatology – for wart removal
3. Medical sample storage

Also, liquid nitrogen can be easier and less expensive to transport in some situations than pressurized, gaseous nitrogen.

3.2.3 Argon

As can be seen from Table 3.1, argon is almost 1 % of the Earth's lower atmosphere. Despite research into determining reaction chemistry that incorporates argon, none has yet successfully isolated an argon compound that is stable at or near ambient temperature (although an ArHF compound has been isolated at −233 °C).

Argon is produced on a large scale annually – over 500,000 tons – and continues to find use in situations where an inert atmosphere is required. Laboratory dry boxes often use argon, but an inert welding blanket is the most widespread consumer end-use. In such cases, an oxy-acetylene torch consists of a jacketing tube through which argon is passed, surrounding the flame, which is some mixture of ignited oxygen and acetylene passing through a central tube. This is so that no oxidation occurs at the spot being welded. For welding reactive metals, such as tungsten or titanium, argon blankets at the surface prevent this over-oxidation.

Argon also finds use as a fire extinguishing medium in select cases in which the equipment to be extinguished is expensive, and minimizing damage to it is important.

3.2.4 Carbon dioxide

Traditionally, carbon dioxide (CO_2) is made when beverages are fermented, although the current, industrial method of producing the material is through gathering carbon dioxide which is often the by-product of many different industrial processes.

The formation of dry ice, the common term for solid carbon dioxide, occurs through the cooling and expansion of gaseous carbon dioxide. The steps to achieve this can be divided as follows:
1. Carbon dioxide is gathered from any large-scale process (which still can include fermentation).
2. The gas is pressurized until it liquefies ($\approx$5.1 atm or greater).
3. The liquid is expanded quickly, which evaporates some, while the heat transfer of the evaporation allows the remainder to freeze.
4. The re-gasified CO_2 is again liquefied and the process repeated.

Dry ice finds a number of industrial uses and some consumer end uses as well. It has been sold as a large-scale commodity since 1924. The first market for dry ice was home refrigeration units. In general, the uses of carbon dioxide, either as a gas or a solid, can be listed as follows:
1. Food preservation, cooling, and storage
2. Production of carbonated beverages
3. Blast cleaning – shooting a stream of compressed air and dry ice pellets at surfaces that need to be cleaned and free of residues

4. As supercritical fluid in enhanced oil recovery operations
5. During wine production – keeps grapes cool and controls the rate of fermentation
6. As an insecticide – usually in large containers, such as grain silos

Additionally, there are numerous small-scale uses, such as fog machines for both theater effects and night clubs or haunted houses, as well as de-gassing tanks which normally hold volatile, flammable chemicals.

Dry ice can be stored for days and at times weeks in conventional freezers, although slow sublimation does occur under such conditions.

3.2.5 Hydrogen

Hydrogen and helium are the two gases that we will address in this chapter, which are not recovered from air liquefaction. Hydrogen is stripped from the light fraction of crude oil, generally methane or ethane in a process called "steam reforming." While this is obviously an organic source for the gas, there are several uses for it that qualify as inorganic. The largest is the production of ammonia, which will be discussed in Chapter 4.

Hydrogen gas is largely produced from two reactions, both of which are shown in Scheme 3.1, each being a different process.

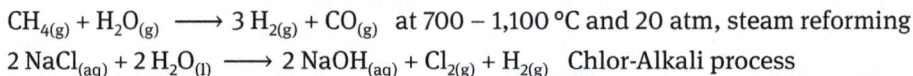

$$CH_{4(g)} + H_2O_{(g)} \longrightarrow 3\,H_{2(g)} + CO_{(g)} \text{ at } 700 - 1{,}100\,°C \text{ and } 20 \text{ atm, steam reforming}$$
$$2\,NaCl_{(aq)} + 2\,H_2O_{(l)} \longrightarrow 2\,NaOH_{(aq)} + Cl_{2(g)} + H_{2(g)} \text{ Chlor-Alkali process}$$

Scheme 3.1: Two major methods of hydrogen production.

The first reaction is the simplified representation of steam reforming. The second is the simplified reaction chemistry for what is called the chlor-alkali process, which will be discussed in more detail in Chapter 7. The economic driver in this case is the production of sodium hydroxide, also called "industrial caustic", as well as chlorine. Hydrogen is sometimes recovered, but is considered a secondary product in this process.

Besides the large-scale use of hydrogen for ammonia production, most hydrogen that is consumed is used in a wide variety of organic reactions, usually in the production of some further chemical. Perhaps obviously, edible fats and oils that have been hydrogenated must use hydrogen to manufacture the final end-product, for example, margarines or other food products.

3.2.6 Helium

Helium is one of only two elements originally found through examination of an extra-terrestrial source, in this case the sun (the other is iron from ancient meteorites). While helium is found in oil deposits, it is not found in the atmosphere in any amount that can be recovered in a cost-efficient manner, being too light to remain in it. It is normally found in uranium and thorium minerals, and in oil and natural gas deposits, as it is the remnant of alpha particle decay. The large-scale recovery and isolation of helium is from the natural gas fraction of gas wells and the lightest fraction of petroleum distillation.

Liquefied helium finds cryogenic uses, and thus is tracked by some national agen-cies, including the USGS Mineral Commodity Summaries in the United States [6]. The USGS states, "There is no substitute for helium in cryogenic applications if temperatures below −429°F are required" [6]. This makes the gas a strategically important material, since it super-cools magnets in magnetic resonance imaging (MRI) instruments.

The distribution of helium throughout the world is shown in Figure 3.1. The total estimated reserve is 160 billion cubic meters [6].

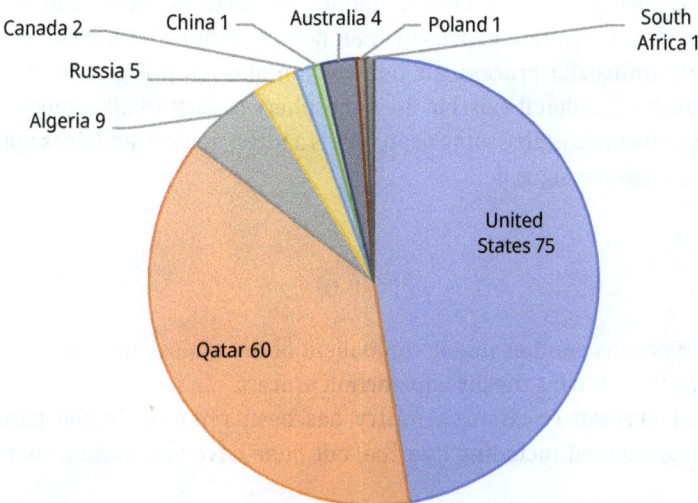

Figure 3.1: World helium reserves, in billions of cubic meters.

There are several uses for helium, and while party balloons probably represent the end use with which most people are familiar, that is actually a rather small portion of the whole. A list of helium uses for both the gas and liquid include:

1. Cooling for magnetic resonance imaging (MRI) and nuclear magnetic resonance (NMR) instruments
2. Pressurizing
3. Purging gas for various air-free systems

4. Semiconductors and fiber optics – cooling as fiber is removed from the furnace [3]
5. Inert atmospheres
6. Welding blankets
7. Detection of leaks in gas piping systems
8. Underwater breathing mixtures, such as trimix and heliox.

There are several other smaller uses for helium as well, such as the carrier gas in GC-MS instruments.

3.2.7 Neon

As shown in Table 3.1, neon represents a small fraction of the air, and can be captured in the air liquefaction process. This was one of the initial offerings of Air Liquide Company, with neon lighting being marketed as a possible source for home lighting as well as for advertising [2].

The major use of neon is indeed in neon lighting, in signs and elsewhere. Putting a voltage through the gas while in an enclosed container produces the now common orange–red glow. Neon is usually more expensive than helium, despite both being co-products of some larger industrial process, air liquefaction and natural gas mining. Neon's expense is tied to the fact that it exists in the atmosphere in such small amounts.

While helium–neon lasers do utilize some neon, this is a niche market, and does not consume nearly as much neon as signs do.

3.2.8 Krypton

As seen in Table 3.1, krypton is another minor component of the atmosphere, but can also be extracted and isolated during the air liquefaction process.

A small amount of krypton reaction chemistry has been reported in the past decades, usually some compound including fluorine, but none have yet found a commercial use.

Overall, krypton has few uses that are large enough that they can be considered industrial, but "neon" signs do often contain krypton. Different amounts of krypton mixed with the neon produce a wider array of colors. Also, krypton can be used for an intense white light source, which is useful in some photographic situations.

3.2.9 Xenon

Like krypton, xenon is a very small component of the atmosphere, and is also separated from liquefied air. But unlike the lighter noble gases, xenon undergoes chemical reactivity and in the past decade has been found to react with fluoride and oxygen. None of the known xenon compounds has yet found a large-scale industrial use.

Some niche applications for xenon do exist, mostly as fill gas in arc lamps and some other lighting. This is because xenon produces a bright white light when properly excited. Quite recently, xenon has been found to be useful as a medical general anesthetic, although existing anesthetic gases continue to claim almost all of this niche application [7].

3.3 Recycling

None of the gases discussed in this chapter are recycled in any large-scale way. When used as elemental gases, all but helium can be reclaimed from air over and over. Because helium is a nonrenewable resource, and because its prices are affected by the prices of oil and natural gas, some large consumers of helium have recently begun to examine methods of collecting helium. Thus, hospitals and universities which use large amounts of liquefied helium for MRI and nuclear magnetic resonance spectroscopy (NMR), respectively, are reaching a point where it is economically feasible to recycle the material.

Bibliography

[1] Almqvist, Ebbe (2003) *History of Industrial Gases*, 2003, ISBN-10: 0387978917.
[2] Air Liquide. Website. (Accessed 19 December 2023, as: https://www.airliquide.com).
[3] Air Products. Website. (Accessed 19 December 2023, as: https://www.airproducts.com, Hydrogen, Hydrogen recovery for Industrial Applications; and Helium, 354-15-009-GLB-secure-and-reliable-helium-supply, pdf).
[4] The Linde Group. Website. (Accessed 19 December 2023, as: https://www.linde.com).
[5] Praxair. Website. (Accessed 19 December 2023, as: https://www.praxair.co.in).
[6] United States Geological Survey, Mineral Commodity Summaries, 2023. Website. (Accessed 18 December 2023 as: https://www.usgs.gov, https://doi.org/10.3133/mcs2023, as a downloadable pdf).
[7] XENON in medical area: emphasis on neuroprotection in hypoxia and anesthesia, Ecem Esencan, Simge Yuksel, Yusuf Berk Tosun, Alexander Robinot, Ihsan Solaroglu and John H Zhang, *Medical Gas Research* 2013, 3:4 doi:10.1186/2045-9912-3-4.

4 Nitrogen-based inorganic compounds

4.1 Introduction

For millennia, farmers have known that plants need some form of fertilizer to help them grow, and to get the maximum amount of food for the seeds they have planted. The major source of fertilizer throughout this time has been animal manure. Farmers also knew that grinding bones from slaughtered animals and plowing this back into the soil was another way to improve the crop yields. The first change in this age-old system came only in the 1800s, when the Chincha Islands off the coast of Peru were mined for the bird guano that had accumulated on them over the course of centuries, as migratory birds stopped at the islands and left guano behind. The islands had been unsettled until their economic potential was discovered. From 1864 to 1866 Spain fought both Peru and Chile in what is called the Chincha Islands War, in an attempt to reassert Spanish control over the islands, because of their overall economic value.

This and other guano mining operations had a brutal and horrific side to it in terms of its human cost, even as the operations brought up crop yields in Europe and North America. Many workers worked the islands under virtually slave labor conditions, with no hope of escaping the islands. Apparently, some chose suicide by jumping off the guano cliffs rather than a slow death from overwork. Historians continue to debate whether some portion of these workers were enslaved or kidnapped and shipped to the islands, or whether all were paid workers doing a very difficult job [1].

The USA actively involved itself in the search for such islands under what was called the Guano Islands Act of 1856 (at the time, guano was used to produce saltpeter for gunpowder, as well as fertilizer). The last United States unincorporated island territory to support such an operation is Navassa Island, off the coast of Haiti, which ceased operations in 1901.

4.2 Ammonia

Synthetic ammonia was pioneered in Germany by chemist Fritz Haber during the First World War. At the outbreak of the war, the political situation was such that the British Royal Navy was able to disrupt commercial shipping to Germany so thoroughly that fertilizer could not be imported. This left Germany in the position that its farmers would not be able to feed the population without some other form of fertilizer. Thus, a method by which artificial fertilizer – artificial ammonia – could be produced was imperative. Fritz Haber answered this challenge by finding a method through which elemental nitrogen and elemental hydrogen could be combined directly.

https://doi.org/10.1515/9783111329512-004

$$N_{2(g)} + 3\,H_{2(g)} \longrightarrow 2\,NH_{3(g)}$$

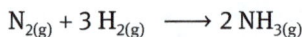

Scheme 4.1: Ammonia synthesis.

The reaction chemistry of ammonia production is very straightforward, and is shown in Scheme 4.1. It is a direct addition – redox reaction in which the two elements combine to form the ammonia product.

The starting materials have been discussed in the previous chapter. Nitrogen comes from the atmosphere, and hydrogen is predominantly sourced from methane, and thus ultimately fossil sources. When Haber originally brought this process to a large scale, the source of hydrogen gas was not methane, but rather coal. The USGS Mineral Commodity Summaries track ammonia as what they call "fixed nitrogen" [2].

Although the reaction chemistry is very simple, the reaction conditions should be examined, and, indeed, are rather extreme. Elevated temperature and pressure are required – 100 atm and 450 °C – as is an iron-based catalyst. Even under these conditions, a single contact does not yield complete conversion to ammonia. This means that the reactants are passed through the system and brought into contact with each other multiple times. Several companies produce ammonia, including the following:
1. CF Industries [3]
2. Dyno Nobel [4]
3. Cornerstone Chemical Company [5] – which also markets urea
4. Dakota Gasification Company [6]
5. KBR [7]

The high percentage of nitrogen in ammonia makes it an excellent form of fertilizer, although there are other forms that are generally sold as solids. Indeed, CF Industries attests at their web site: "Anhydrous ammonia, which contains 82 percent nitrogen, is the most concentrated nitrogen fertilizer. Because of its high nitrogen content, it is often the most cost-effective nitrogen fertilizer [3]". Additionally, it can be transported in numerous ways, including tank trucks. Figure 4.1 shows such a tank truck, with arows pointing both to the ammonia label and the fact that it is an inhalation hazard.

4.3 Ammonium nitrate

Ammonium nitrate is another major industrial chemical that is used primarily for fertilizer. Like ammonia, it delivers a significant amount of nitrogen to the soil and the plant. Unlike ammonia, it is a crystalline solid, which is useful when one considers the ease of handling of a solid as opposed to that of a liquid. Its high solubility in water is another useful property, once it has been delivered to farm fields.

Figure 4.1: Rear view of an ammonia truck.

The reaction chemistry for ammonium nitrate production is shown in Scheme 4.2.

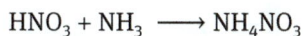

$$HNO_3 + NH_3 \longrightarrow NH_4NO_3$$

Scheme 4.2: Ammonium nitrate synthesis.

The chemistry is a simple addition reaction which utilizes both ammonia and nitric acid. It is an acid-base addition, is very exothermic, and is run in water. The product is often concentrated by a final evaporation of excess water. Depending on how much water is evaporated, the final product can be over 99 % pure. Reaction conditions can vary, depending on whether the product is to be formed as prills or as pellets of crystalline material [8]. In both cases, the finished product must be kept dry until use, because it has the ability to absorb moisture from the air, and form larger, solid chunks which are difficult to work with.

4.4 Nitric acid

Nitric acid has been produced by what is called the Ostwald process for over a century. The process was patented in 1902 [9]. Ammonia becomes one of the starting materials for the process, with oxygen and water being the other two major reactants. In terms of reaction chemistry, the synthesis of nitric acid can be represented by the following, shown in Scheme 4.3.

$$5 O_{2(g)} + 4 NH_3 \longrightarrow 4 NO_{(g)} + 6 H_2O_{(g)}$$

followed by

$$O_{2(g)} + 2 NO_{(g)} \longrightarrow 2 NO_{2(g)}$$

then

$$3 NO_{2(g)} + H_2O_{(l)} \longrightarrow NO_{(g)} + 2 HNO_{3(aq)}$$

and finally

$$O_{2(g)} + 4 NO_{2(g)} + 2 H_2O_{(l)} \longrightarrow 4 HNO_{3(l)}$$

Scheme 4.3: Ostwald process chemistry for nitric acid synthesis.

Each of the reactions is exothermic, with the first having $\Delta H = -905$ kJ, the second $\Delta H = -114$ kJ, and the third $\Delta H = -117$ kJ. The ammonia oxidation in the first step still requires a catalyst, usually rhodium or platinum, and elevated pressure (roughly 9 bar) and temperature (roughly 500 K or 220 °C). The nitrogen monoxide co-produced with nitric acid is fed back into the system to maximize the ultimate production.

Despite the above-mentioned purity levels, nitric acid can be concentrated to as high as 98 % when sulfuric acid is used to dehydrate it. At this level of concentration, the product is called "fuming nitric acid."

The Ostwald process does not produce nitric acid in high purity, at least when compared to other bulk chemicals. Routinely, nitric acid is made in commercial grades from 52 to 68 % purity.

There are numerous manufacturers for nitric acid, including:
1. Dow Chemical [10]
2. Potash Corporation [11]
3. CF Industries [12]

This list is not all inclusive.

Almost all nitric acid is used in the production of ammonium nitrate fertilizer, as shown above. This equates to tens of millions of tons annually. It does have some crossover uses in organic chemistry as well, since the introduction of multiple nitro-groups onto organic compounds tends to make them explosive. Multiply nitrating toluene produces the famous TNT (more properly, 2,4,6-trinitrotoulene). Several other explosives also have multiple nitro groups attached to an organic molecule.

4.5 Urea

Urea, an organic material first isolated from animal urine, is another nitrogen-containing compound which is now produced commercially in what can be considered to be an inorganic process as well as an organic one. It exists as a white, crystalline solid. One of the first prominent chemists to work on urea was Friedrich Wöhler, who synthesized it from an inorganic source in 1828. This was unexpected at the time, as the belief then was strong that only living material could make such compounds, and that inorganic material could never make some product associated with any living material. Scheme 4.4 shows the reaction chemistry by which urea is produced.

$$CO_2 + 2\,NH_3 \longrightarrow NH_2COO^-NH_4^+ \longrightarrow NH_2CONH_2 + H_2O$$

Scheme 4.4: Urea production.

The large-scale production of urea is done by the Bosch–Meiser urea process, which has been the major process since the 1920s.

More than 90 % of urea finds use in producing fertilizers. Like ammonia, it has a high nitrogen content. Like ammonium nitrate, it is a white solid, and thus somewhat easier to use than a liquid.

When tracking ammonia and the products produced from it, the USGS Mineral Commodity Summaries states, "Urea, ammonium nitrate, nitric acid, ammonium phosphates, and ammonium sulfate were, in descending order of quantity produced, the major derivatives of ammonia produced in the United States" [2]. The Fertilizer Institute also notes the importance of urea as a fertilizer and as a way to enhance plant growth in agriculture [13].

4.6 Ammonium sulfate

Ammonium sulfate is another nitrogen compound made in large-scale quantities that is primarily used as a fertilizer. It finds its major use on slightly alkaline soils, since when it is applied, the ammonium is released into the soil and forms small but significant amounts of acid, thus lowering the overall pH of the soil.

Scheme 4.5 shows the reaction chemistry for the production of ammonium sulfate in a simplified form.

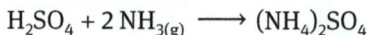

$$H_2SO_4 + 2\,NH_{3(g)} \longrightarrow (NH_4)_2SO_4$$

Scheme 4.5: Ammonium sulfate production.

The reaction is run by mixing water vapor and gaseous ammonia, and introducing this to a saturated solution of ammonium sulfate that also contains free sulfuric acid. The resulting addition reaction is exothermic, and on a large scale, the process is usually run at 60 °C. Periodically, more sulfuric acid must be added to continue the reaction.

As a crystalline material, a reaction chamber filled with ammonia gas can have gaseous sulfuric acid introduced to it. The reaction chemistry for this is essentially the same as shown in Scheme 4.4. In this form, the heat of the reaction drives off any water in the system, allowing the collection of a dry product.

A lesser amount of ammonium sulfate is used as a food additive, often to adjust the pH (the acidity level) of foods such as bread. It has been assigned the European Union designator E517, and within the USA has been noted to be, "generally recognized as safe" (GRAS) by the US Food and Drug Administration.

4.7 Hydrazine

Hydrazine, H_2NNH_2, is one important nitrogen compound produced on a large scale from ammonia that is not used in the production of fertilizers. Rather, most hydrazine is used to prepare a variety of organic compounds that require the amine group or some other nitrogen-containing group. The majority of hydrazine that is manufactured is used in blowing agents for the further manufacture of plastics and plastic foams. There are several smaller applications as well, including its use as a starting material for the gas produced in some automobile air bags.

There are several methods by which hydrazine is produced. Scheme 4.6 shows what is called the Olin–Raschig process, which has been used for over a century.

$$NH_2Cl + NH_3 \longrightarrow HCl + H_2NNH_2$$

Scheme 4.6: Hydrazine production.

The chloroamine is formed by the reaction of an excess of ammonia with sodium hypochlorite (NaOCl). This process also depends upon using a large excess of ammonia, and temperatures of approximately 125 °C. Sodium chloride does form as a by-product, and must be removed. Furthermore, any water in the final product mixture must be distilled away.

4.8 Nitrogen pollution

The problem persists of pollution caused by an excess of nitrogen-based fertilizers finding their way into major waterways. Fertilizer is obviously necessary to produce the amount of food required to feed the world's current population. But the application of too much fertilizer produces what is referred to as non-point source pollution. Excess fertilizer is washed from fields by rainfall, where it makes its way into small streams and rivers, which eventually flow into larger rivers, which eventually flow into the oceans.

The Gulf of Mexico has, for example, an area that encompasses the estuary of the Mississippi River which is often called a dead zone, because of the total pollution in the water there. The United States Environmental Protection Agency comments: "The lack of oxygen makes it impossible for aquatic life to survive. The largest dead zone in the United States – about 6,500 square miles – is in the Gulf of Mexico and occurs every summer as a result of nutrient pollution from the Mississippi River Basin" [14]. A contribution to this total comes from fertilizer, as it does at the mouths of the world's other large rivers. The long-term solution to this problem appears to be using fertilizers carefully, with applications being as site-specific as possible.

4.9 Recycling

For all of the products discussed in this chapter, the materials are delivered to some end user, often farmers, and thus there is no recycling that is possible. Rather, efforts are directed at the controlled application of fertilizers, to minimize waste.

In addition, the prices for the materials discussed here have remained very low for decades, and thus there is little financial incentive for any large recycling effort.

Bibliography

[1]   *The Great Guano Rush: Entrepreneurs and American Overseas Expansion*, J.M. Skaggs, 1994, ISBN: 0-312-10316-6.

[2]   United States Geological Survey, Mineral Commodity Summaries, 2023. Website. (Accessed 18 December 2023 as: https://www.usgs.gov, https://doi.org/10.3133/mcs2023, as a downloadable pdf).

[3]   CF Industries. Website. (Accessed 19 December 2023, as: https://www.cfindustries.com/ammonia).

[4]   Dyno Nobel. Website. (Accessed 19 December 2023, as: https://www.dynonobel.com, Product Hub).

[5]   Cornerstone Chemical Company. Website. (Accessed 19 December 2023, as: https://cornerstonechemco.com).

[6]   Dakota Gasification Company. Website. (Accessed 15 December 2014, as: https://www.dakotagas.com/products/fertilizers/index).

[7]   KBR. Website. (Accessed 19 December 2023, as: https://www.kbr.com/en-gb/node/2356).

[8]   ANNA, Ammonium Nitrate Nitric Acid Producers Group. Website. (Accessed 19 December 2023, as: https://an-na.org).

[9]   Ostwald process patent, GB 190200698, Ostwald, Wilhelm, "Improvements in the Manufacture of Nitric Acid and Nitrogen Oxides", published January 9, 1902, issued March 20, 1902.

[10] Dow Chemical. Website. (Accessed 19 December 2023, as: https://www.dow.com).

[11] American Potash Corp. Website. (Accessed 19 December 2023, as: https://americanpotash.com).

[12] CF Industries. Website. (Accessed 19 December 2023, as: https://www.cfindustries.com/products/other).

[13] The Fertilizer Institute. Website. (Accessed 19 December 2023, as: https://www.tfi.org/policy-center/safety-security).

[14] US EPA. Website. (Accessed 19 December 2023, as: https://www.epa.gov. The Effects: Dead Zones and Harmful Algal Blooms. https://www.epa.gov/nutrientpollution/).

5 Fertilizers

5.1 Introduction

The different components within fertilizers come from a wide variety of sources; several of them are minerals. Modern fertilizers can be adjusted and designed for different soil and plant types, but almost always include nitrogen, phosphorus, and potassium. The addition of synthetic fertilizers to farm fields throughout the 20th century has enabled the production of more food than ever before, and has occurred in conjunction with the biggest increase in human population in recorded history. Numerous institutions exist to promote the development and sales of fertilizers, including those centered in Europe, North America, and Australia [1–3].

Figure 5.1 shows the growth of human population versus the course of time, from the early 19th century through the present.

The vertical axis is the number of people in billions, and the horizontal axis is listed in years. It can be seen that it took almost 125 years for the population to double from 1 billion to 2 billion people, in 1927. The 8 billion mark that is listed at the year 2026 is obviously speculative, but the information has been compiled by the United Nations and the United States Census Bureau, and shows the expected increase in population [4, 5]. The explosive growth seen between the year 1925 and the present day correlates quite well with the growth in the fertilizer industry. Perhaps obviously, other factors also figure into this population increase, such as the production and use of antibiotic medicines starting in 1938, but the use of synthetic fertilizers does correspond closely with the rise in human population.

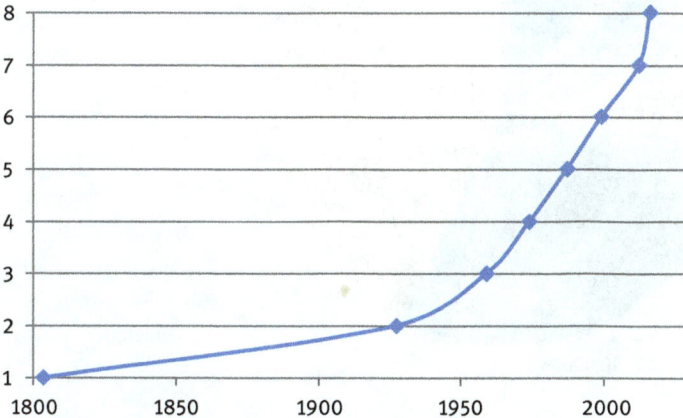

Figure 5.1: World population growth, in billions.

https://doi.org/10.1515/9783111329512-005

5.2 Nitrogen based

We have discussed nitrogen-based fertilizers in the previous chapter, but will restate the following: ammonia can be added to soil directly or can be put into solution first, then added to farm fields. The sale of ammonia and ammonia-derived fertilizers has become one of the largest chemical processes in the world, rivaled only by the production of sulfuric acid (which itself finds applications in fertilizer production).

5.3 Phosphorus based

Phosphorus in fertilizers comes from phosphate-containing minerals. While there are several of them, fluorapatite, $Ca_5(PO_4)_3F$ is the most often mined. Hydroxyapatite, $Ca_5(PO_4)_3OH$ is also widely used as a starting material. Because the production of fertilizer is so vital, the USGS Mineral Commodity Summaries tracks phosphate rock production, and records it in thousands of metric tons [6, 7]. Figure 5.2 shows the breakdown by nation of such production.

It is apparent from Figure 5.2 that phosphate rock is mined extensively throughout the world, although China, the USA, and Morocco currently dominate production.

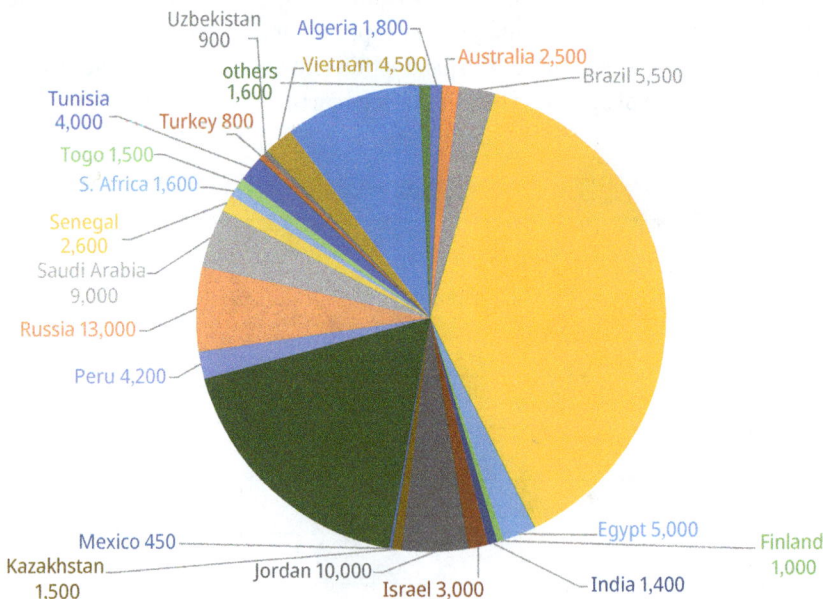

Figure 5.2: Phosphate rock production (in thousands of metric tons).

Reacting such materials with sulfuric acid produces phosphoric acid. As mentioned in Chapter 2, this becomes a major use for sulfuric acid. The general reaction chemistry is shown in Scheme 5.1.

$$2\,Ca_5(PO_4)_3F + 10\,H_2SO_4 + 20\,H_2O \longrightarrow 18\,H_3PO_4 + 2HF + 10\,CaSO_4 \cdot 2H_2O$$

Scheme 5.1: Phosphoric acid production.

This process also produces an enormous amount of calcium sulfate which precipitates out of the solution. Incidentally, some calcium sulfate finds use in foods such as bread, beer, cheese, and tofu products, where it keeps bread dough from being sticky, and functions as a preservative.

The best means of delivering phosphorus to the plant in any fertilizer is to do so in as soluble a form as possible. This is often done by the reaction of phosphoric acid with potassium hydroxide, which yields the highly soluble dipotassium hydrogen phosphate. The reaction chemistry for this is shown in Scheme 5.2.

$$H_3PO_4 + 2\,KOH \longrightarrow K_2HPO_4 + 2\,H_2O$$

Scheme 5.2: Dipotassium hydrogen phosphate production.

Interestingly, dipotassium phosphate is also used as a food additive, but is done so in much smaller quantities than that which is used for fertilizer. The European number for the three possible potassium phosphates is E340. All are considered antioxidants.

5.4 Potassium-containing

Potash is a potassium-containing material that is produced on a large scale through traditional mining operations as well as what are called solution mining operations. Traditional mining of potassium minerals is no different than mining any other ores – the overburden is removed and the potash is extracted. Solution mining takes advantage of the high solubility of potassium salts in water. Piping systems are inserted into the earth down to the potassium minerals and hot water is injected down the pipes. The resulting solution is piped out and allowed to evaporate in pools in the sun. The resulting solid may need further treatment to separate KCl, one of the main components of potash, from NaCl. This is usually done by adding proprietary amines to the pools, which react preferentially with the potassium chloride, leaving sodium chloride as an insoluble material.

Table 5.1: Global Potash Manufacturers.

Company Name	National HQ	Comments
Agrium, Inc.	Canada	Also markets wide array of ammonia-based products
Arab Potash Company	Jordan	
Belaruskali	Belarus	
BHP Billiton, Ltd.	Australia	
Boulby Mine, ICL Group	Britain	
Compass Minerals International, Inc.	USA	
Dead Sea Works	Israel	Also produces bromine
Elemental Minerals	Australia	
Encanto Potash Corp.	Canada	
Great Quest Metals, Ltd.	Canada	Developing phosphate mining in Mali
IC Potash Corp.	USA	
Intrepid Potash, Inc.	USA	
K+S AG	Germany	Claims to be world's largest salt producer
Karnalyte Resources, Inc.	USA	
Lithium Americas Corp.	Canada	Also major producer of lithium carbonate
MBAC Fertilizer Corp.	Canada	
The Mosaic Company	USA	Also produces concentrated phosphate
Orocobre Limited	Australia	Also produces lithium and borax
Phoscan Chemical Corp.	Canada	
Potash Corp. of Saskatchewan	Canada	
Qinghai Salt Lake Potash Company	China	
Red Metal, Ltd.	Australia	Also mines metals such as uranium
Rio Tinto Ltd.	Australia	Rio Tinto group has widespread operations on 6 continents
Sociedad Quimica y Minera	Chile	Also produces lithium and iodine from salt brines
South Boulder Mines, Ltd.	Australia	
Uralkali	Russia	World's largest single producer
Verde Potash Plc	USA	
Western Potash Corp.	USA	

Potash production is enormous. Over 30 million metric tons are mined annually, and the USGS Mineral Commodity Summaries estimates over 9.5 billion tons as a global reserve [7]. Major producers of potash are listed in Table 5.1.

5.5 Mixed fertilizers and the NPK rating system

As the name suggests, mixed fertilizers are formulated to have a combination of nitrogen, phosphorus, and potassium in them. What is called the "NPK" system is a way to rate the amounts of nitrogen, phosphorus, and potassium in them, in that order. Thus, a number such as 10–6–4 means the material has 10 % elemental nitrogen, 6 % elemental phosphorus, and 4 % elemental potassium.

Mixed fertilizers are useful for delivering multiple nutrients to the soil in one dose, but can be formulated so that the release is not immediate. When the solid fertilizer is combined with a degradable, generally water-soluble resin, such as urea-formaldehyde, the material dissolves in rainwater or irrigation water more slowly, and thus is taken up into the soil and the plant over a longer period of time. Sulfur can also be mixed or coated onto solid fertilizer particles to delay immediate uptake. The use of sulfur can also be beneficial to certain soils.

5.6 Recycling and re-use

There are no programs for the recycling of any fertilizers, as all are considered consumer end-use products. As mentioned in Chapter 4 in Section 4.7, application of excess fertilizer causes water-borne pollution. Significant effort and resources continue to go into the science and practice of delivering just the right amount of fertilizers to specific soils and fields [1–3, 8–10].

Bibliography

[1]   CropLife Europe. Website. (Accessed 19 December 2023, as: https://croplifeeurope.eu).
[2]   The Fertilizer Institute. Website. (Accessed 19 December 2023, as: https://www.tfi.org/).
[3]   Fertilizer Australia. Website. (Accessed 19 December 2023, as: https://fertilizer.org.au).
[4]   United Nations, The World at Six Billion. Website. (Accessed 19 December 2023, as: https://www.un.org/development/desa/pd/sites/www.un.org.development.desa.pd/files/documents/2020/Jan/un_1999_6billion.pdf).
[5]   United States Census Bureau. Website. (Accessed 19 December 2023, as: https://www.census.gov/popclock).
[6]   International Fertilizer Association. Website. (Accessed 19 December 2023, as: https://www.fertilizer.org).
[7]   United States Geological Survey, Mineral Commodity Summaries, 2023. Website. (Accessed 18 December 2023 as: https://www.usgs.gov, https://doi.org/10.3133/mcs2023, as a downloadable pdf).
[8]   Fertilizer Canada. Website. (Accessed 19 December 2023, as: https://fertilizercanada.ca).
[9]   United Nations Environmental Programme. Website. (Accessed 19 December 2023, as: https://www.unfoundation.org/?s=fertilizer).
[10] Food and Agriculture Organization of the United Nations. Website. (Accessed 19 December 2023, as: http://www.fao.org/home/en).

6 Calcium and limestone-based products

6.1 Introduction

Limestone – calcium carbonate, $CaCO_3$ – has an ancient history as a building material, having seen use in the buildings of many cultures. It is mined in many places throughout the world, has become a major industry product, and in many deposits is essentially the fossilized remains of prehistoric sea life. The finest forms of limestone, the most visually attractive, are the different forms of marble, which have been used for statuary and other decorative stonework, also since ancient times. But even single crystals of calcium carbonate – sometimes called calcite – can be attractive, as shown in Figure 6.1. All forms of it are used so widely today that there are several trade organizations devoted to it [1–3].

Figure 6.1: Calcium carbonate crystal.

Because limestone and the materials made from it have been used for centuries, a rather imprecise nomenclature has built up around it. Table 6.1 lists the older, established names of these materials, as well as their chemical formula.

Table 6.1: Names of materials derived from limestone.

Formula	Chemical Name	Common Names
$CaCO_3$	Calcium carbonate	Limestone, marble, chalk, calcite
CaO	Calcium oxide	Lime, unslaked lime, quicklime
$Ca(OH)_2$	Calcium hydroxide	Slaked lime, hydrated lime
Na_2CO_3	Sodium carbonate	Soda ash, washing soda, soda crystals

https://doi.org/10.1515/9783111329512-006

6.2 Lime

6.2.1 Lime production

Lime is produced by the direct heating of limestone, liberating the carbon dioxide and producing calcium oxide. The reaction chemistry is remarkably simple, as shown in Scheme 6.1.

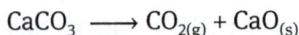

$$CaCO_3 \longrightarrow CO_{2(g)} + CaO_{(s)}$$

Scheme 6.1: Lime production.

This reaction is one of the largest chemical processes performed annually, with hundreds of millions of tons produced each year. The simplicity of the reaction hides the fact that it is highly energy intensive, as it is run at approximately 1,200 °C. The USGS Mineral Commodity Summaries tracks the production of lime worldwide, and Figure 6.2 shows the breakdown by country [4].

It is extremely important to note that for reasons of clarity of the figure, this figure excludes the production number for China, which alone produces 310,000 thousand

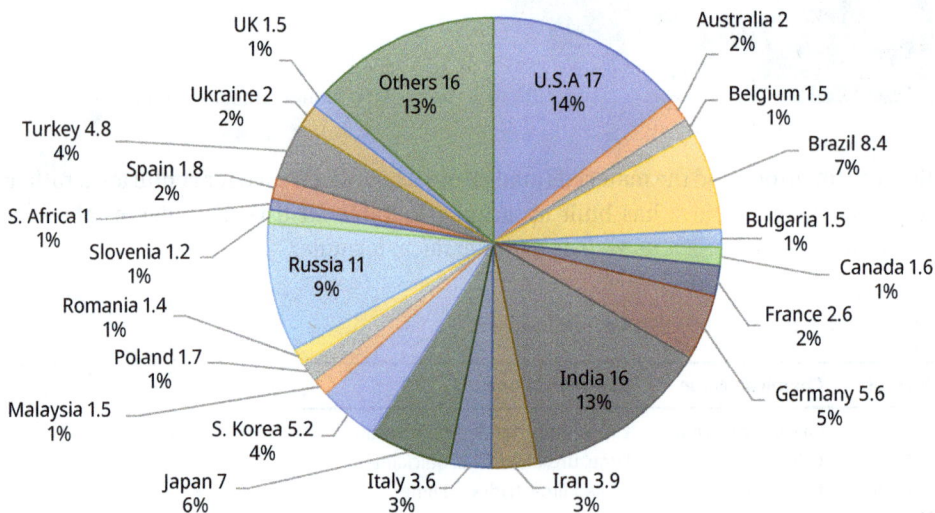

Figure 6.2: Lime production, in thousands of metric tons.

metric tons annually – making China's output greater than the output of all the countries listed in Figure 6.2.

6.2.2 Lime uses

The chapter began with the comment that limestone has been an important building material for centuries. The use of lime in construction is also important, and very widespread as Figure 6.2 shows, but the USGS Mineral Commodity Summaries states that: "Major markets for lime were, in descending order of consumption, steelmaking, chemical and industrial applications (such as the manufacture of fertilizer, glass, paper and pulp, and precipitated calcium carbonate, and in sugar refining), flue gas treatment, construction, water treatment, and nonferrous-metal mining" [4]. Iron and steel production is discussed in Chapter 11, but clearly the two industries and the production of these two commodities are tightly linked.

A simplified version of how lime functions in the production of iron is shown in Scheme 6.2. The silica represents the impurities within iron ores. This is what reacts with lime in a blast furnace to form slag, which is separated from the reduced iron based on its lighter density [5].

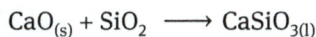

$$CaO_{(s)} + SiO_2 \longrightarrow CaSiO_{3(l)}$$

Scheme 6.2: Slag production from lime.

6.3 Sodium carbonate (or: soda ash)

The Solvay process for the production of sodium carbonate – traditionally known as soda ash because of its sources prior to the development of this process – is the major use of calcium carbonate, and one of the major uses of common salt, sodium chloride. It was scaled up to the industrial level in the 1860s, and thus is a very mature industry today. The reaction chemistry for it involves recycling a series of intermediate materials, but it must always have calcium carbonate added to it, as well as sodium chloride. Figure 6.3 shows the Solvay process, inclusive of the recycling of materials.

The reaction chemistry for the Solvay process can be simplified to a net reaction, as shown in Scheme 6.3. This simplification ignores the ammonia and carbon dioxide, but does show the two products, the sodium carbonate and the calcium chloride [6,7].

The Solvay process re-uses all materials but the calcium chloride, so there is little ammonia that needs to be introduced to the system. Likewise, carbon dioxide is produced by the calcination of the original calcium carbonate and introduced in that way.

$$CaCO_3 + 2\ NaCl \longrightarrow Na_2CO_3 + CaCl_2$$

Scheme 6.3: Solvay process net reaction.

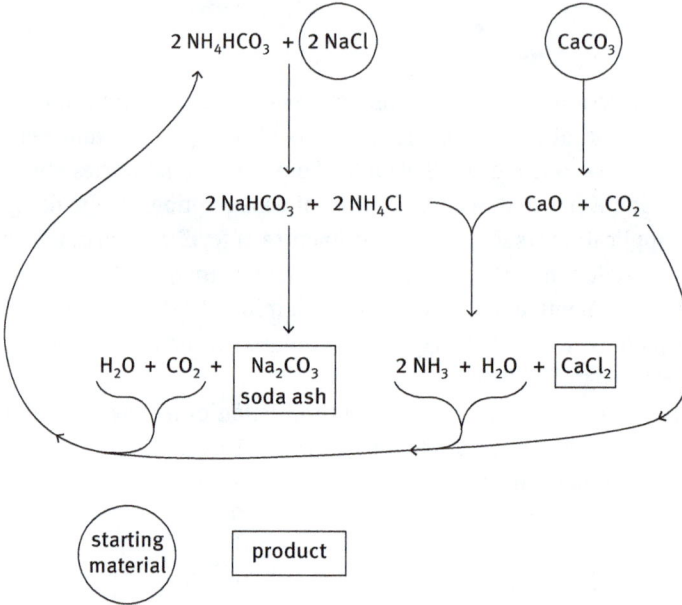

Figure 6.3: The Solvay process.

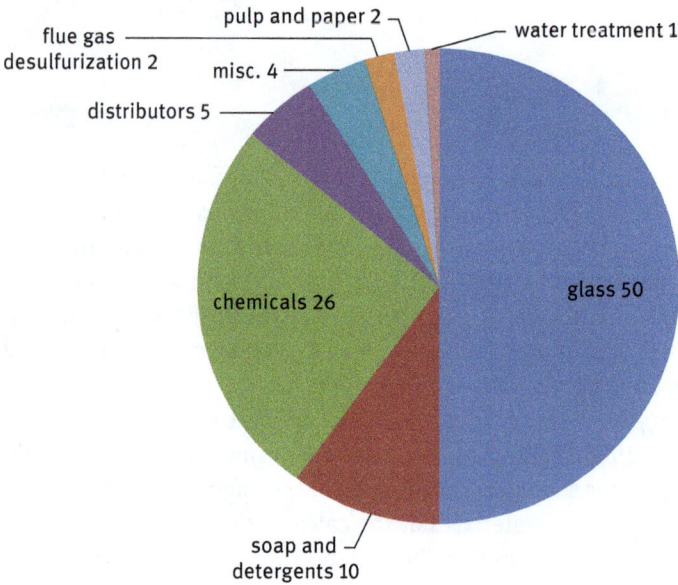

Figure 6.4: Uses of soda ash.

The uses of soda ash span a range of applications. Figure 6.4 shows the major uses, broken down in percentages [8, 9]. Glass will be discussed along with ceramics in Chapter 19.

6.4 Calcium chloride

This is the only material produced in the Solvay process besides the major product sodium carbonate, that is not recycled into the process for further use. It now has found a major use as road salt for de-icing roads in the winter.

Other uses of calcium chloride include: brine for refrigeration, a dehumidifier in enclosed environments, drying tube dessicant, and additive to adjust hardness in swimming pool water.

Calcium chloride has the food additive designation E509 in Europe, it meets the US Food and Drug Administration standard of being generally recognized as safe (GRAS). It finds use in the following foods [10]:

1. A firming agent in canned vegetables
2. A firming agent in soy bean curd
3. A food-grade dessicant
4. Milk additive – adjust calcium content
5. Beer additive – adjust ion content
6. Sport drink electrolyte
7. Pickle flavoring salt

One other use of calcium chloride is its addition to sodium chloride in the production of metallic sodium. Its addition lowers the working temperature of the molten mix.

6.5 Limestone-based construction materials

As mentioned at the beginning of the chapter, limestone has been used as a building material for millennia. North America has large deposits. Bloomington, Indiana, in the US, as well as Kingston, Ontario, in Canada are both known for the amounts of limestone they have produced and used in construction. Also, Europe has a long tradition of building with limestone. Many famous churches and cathedrals have been constructed with limestone, since it was easy to obtain and to carve and shape.

Today limestone continues to find use in a variety of construction applications. Limestone that is not the absolute highest grade is still used as walls and supporting material. The higher grade material – meaning the various grades of marble – are used as decorative facing for walls, floors, and sometimes ceilings. When crushed into small, uniform size pieces, limestone can also be used as base material in roads.

6.6 Recycling

Depending on the grade of limestone or marble, these materials can easily be recycled. Since they are used as building materials, and are not expected to be consumed quickly, there are no formal recycling programs for them. Rather, material is re-used on a case-by-case basis.

The Solvay process recycles almost all of the materials that are not the desired products. There are some calcium and magnesium impurities in any batch fed into the system, and these precipitate out, usually as carbonates. Thus, these become a secondary or waste product for this process.

Bibliography

[1] National Stone Sand and Gravel Association. Website. (Accessed 19 December 2023, as: https://www. nssga.org).

[2] Limestone Association of Australia, Inc. Website. (Accessed 19 December 2023, as: https://limestone. asn.au)

[3] Natural Stone Institute. Website. (Accessed 19 December 2023, as: naturalstoneinstitute.org/about/).

[4] United States Geological Survey, Mineral Commodity Summaries, 2023. Website. (Accessed 18 December 2023 as: https://www.usgs.gov, https://doi.org/10.3133/mcs2023, as a downloadable pdf).

[5] MPA Lime. Website. (Accessed 19 December 2023, as: https://mpalime.org).

[6] Solvay S.A. Website. (Accessed 19 December 2023, as: https://www.solvay.com).

[7] Solvay. Website. (Accessed 19 December 2023, as: solvay.com/en/brands/soda-solvay).

[8] ANSAC. American Natural Soda Ash Corporation. Website. (Accessed 19 December 2023, https://www. ansac.com).

[9] ESAPA – European Soda Ash Producers Association. Website. (Accessed 19 December 2023, as: https://specialty-chemicals.eu/esapa-european-soda-ash-producers-association).

[10] Nedmag B.V. Website. (Accessed 19 December 2023, as: https://www.nedmag.com/markets-and-appliications/food-health).

7 Sodium chloride

7.1 Introduction

Salt has been mined, gathered, or evaporated by people throughout the world for thousands of years. Since ancient times, peoples throughout the world have realized that salt is essential for life. It is well known as a preservative and as a flavor enhancer for food. At times, salt was so valuable that it was used as a means of payment, for example, to Roman troops. The word "salary" and the expression to be "worth one's salt", come from this use of it. There are numerous other expressions still in our language today that relate to salt, and thus testify to its importance in some facet of life. Expressions such as: "taken with a grain of salt", or to be "the salt of the earth", or "to salt the earth behind you", have their roots in antiquity and in the value and uses of this material. Almost all peoples have had salt available to them, although some peoples and some nations that are landlocked have had to trade for salt.

Salt is also one of only a very few chemicals that qualify as being the pillars of modern industrial chemistry. Oil, natural gas, water, air, sulfur, iron, and aluminum are arguably the others. In Chapter 6, we saw that sodium chloride was necessary in the Solvay process. That is only one use of its uses as a chemical feedstock, though. Here we will see its direct uses, as well as its uses in the production of several other commodity chemicals, the largest of which is sodium hydroxide, or "industrial caustic." Overall, salt is used on a large enough scale that there are trade organizations devoted to it [1–3].

7.2 Sodium chloride recovery and production

Sodium chloride is produced either through evaporation of brine solutions, or through various mining methods. Traditional underground mining remains one major means by which salt is produced, although solution mining – injecting hot water into deposits, then extracting the hot brine solution – has become more common in the recent past.

7.2.1 Production and recovery methods

The USGS Mineral Commodity Summaries delineate the means by which salt is produced as follows: salt in brine, rock salt, vacuum pan, and solar salt. Major salt producing companies tend to categorize salt production in the following three ways:
1. Rock salt – Salt is produced from underground deposits. Its purity may be lower than that obtained by other methods, at least until such impurities have been removed through solvation and selective precipitation. This type of salt is still called halite, at least when it is directly removed from underground deposits.

https://doi.org/10.1515/9783111329512-007

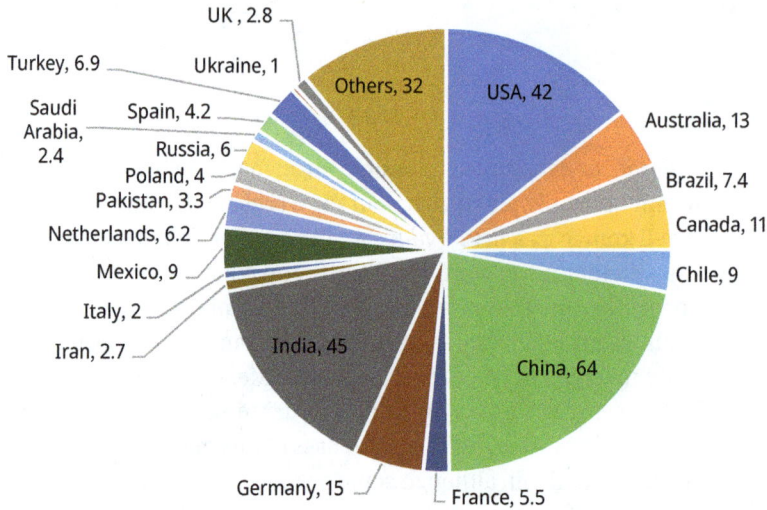

Figure 7.1: Salt production, in millions of metric tons [4].

2. Vacuum pan – Brine solutions are heated in pans while under a partial vacuum. The vacuum lowers the boiling point and speeds evaporation [5, 6]. This tends to produce salt in high purity, and is often combined with solution mining, in which hot water is first injected into the underground deposit.

3. Solar salt – This is considered the oldest method by which salt is concentrated, and is sometimes called the Grainer process. Salt is left in the sun in shallow pools so that the water will evaporate [5, 6]. Ions such as calcium and magnesium can be removed via selective precipitation during the evaporation process. Bromine can also be harvested from salt brines as the evaporation process takes place.

7.2.2 Direct uses for salt

Salt certainly finds numerous chemical uses, in which it is converted into other useful, valuable chemicals. But there are several uses for salt itself. The USGS Mineral Commodity Summaries states: "Highway deicing accounted for about 42 % of total salt consumed. The chemical industry accounted for about 39 % of total salt sales, with salt in brine accounting for 91 % of the salt used for chemical feedstock. Chlorine and caustic soda manufacturers were the main consumers within the chemical industry. The remaining markets for salt were distributors, 9 %; food processing, 4 %; agricultural, 3 %; general industrial, 2 %; and primary water treatment, 1 %" [4]. The amount of salt used for road de-icing competes with calcium chloride, which was discussed in Chapter 6, as well as materials such as wood ashes. Additionally, road salt usage depends upon the severity of winter in the northern hemisphere. For example, the winter of 2012–2013 was relatively mild, which depressed the sales of salt, while that of 2013–2014 was exceptionally cold and snowy across much of northern North America, which increased sales.

7.3 Major chemicals produced from salt

There are numerous chemicals produced from salt, all of which can be divided into two large classes: those being hydro-based processes (requiring water), and those which are pyro-metallurgical (high temperature) in design. Those we discuss in this section are the largest chemical processes that in some way involve salt. There are others, as well.

7.3.1 Sodium hydroxide

The production of sodium hydroxide, called the chlor-alkali process, is a very large scale process, and one that has seen numerous refinements over the past decades. While this is a mature industry, the continued demand for the three products of the process means that design improvements are still being sought. The reaction chemistry for this is not difficult, and is shown in Scheme 7.1.

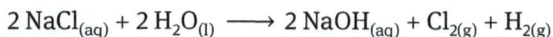

$$2\,NaCl_{(aq)} + 2\,H_2O_{(l)} \longrightarrow 2\,NaOH_{(aq)} + Cl_{2(g)} + H_{2(g)}$$

Scheme 7.1: The chlor-alkali process.

What the reaction does not show is the energy requirement to make this process proceed. The cost of the energy needed to run a chlor-alkali plant is significant, and it is factored into the production costs. There are three broad methods for producing sodium hydroxide, but all produce the same three materials. A general diagram of the chlor-alkali process is shown in Figure 7.2.

The mercury cell
Mercury cell setups for the chlor-alkali process use a flowing layer of elemental mercury as the cathode. This actually amalgamates the sodium, which then must be oxidized back to the ion, after the chlorine has been oxidized to elemental chlorine gas and captured in a separate chamber. The concurrent splitting of water provides the hydroxide ions that combine with the sodium, while elemental hydrogen gas is taken off as a third product.

This operational setup produces sodium hydroxide solution that is as concentrated as 50 %, which is an advantage. The disadvantage of this type of chlor-alkali setup is the large amount of mercury that must be employed. Mercury toxicity is well known, but when such plants operate efficiently, mercury emissions are minimized.

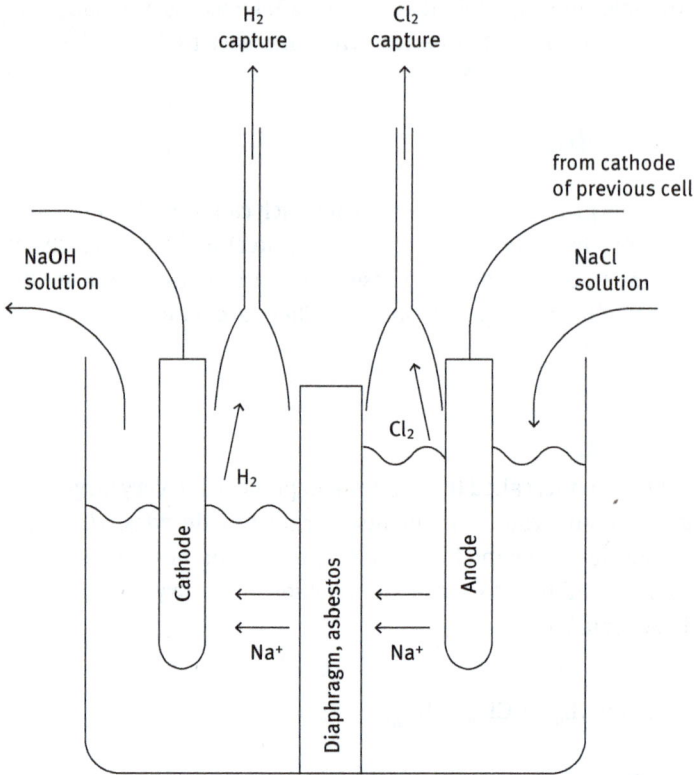

Figure 7.2: General diagram of chlor-alkali process.

The diaphragm cell

The key to this form of chlor-alkali operation is the use of asbestos–Teflon membranes that separate the cathode and anode. This coats the cathode in each cell (cells are linked) and allows effective control of the migration of ions in the electrolyte solution.

The membrane cell

The membrane cell operation of a chlor-alkali process plant replaces the asbestos–Teflon membrane with an ion exchange membrane in each cell. The percentage of plants that operate using this design continues to increase. Its main advantage is the elimination of asbestos from the operation.

There are several major uses for sodium hydroxide, which are tracked by the World Chlorine Council [7]. These uses become a series of cross-over points between what is often considered inorganic and what is considered organic. A nonexhaustive listing is shown in Table 7.1, which includes such cross-over uses as the production of sodium surfactants, as well as monosodium glutamate, a food additive.

Table 7.1: Uses for sodium hydroxide.

NaOH Derivative	End Use Product	Comments
Acrylonitrile	Plastic parts, ABS resins	
Monosodium glutamate	Food additive, flavor enhancer	Enhances meaty flavors of foods
Sodium chlorite	Bleach for textiles	
Sodium cyanide	Adiponitrile, nylon	Also used for gold extraction in mining
Sodium formate	Rare earth element processing	
Sodium lauryl sulfate	Food additive	Emulsifier for egg whites, wetting agent
Sodium stearate	Cosmetics and soaps	Gelling agent

7.3.2 Sodium metal and chlorine

The large-scale production of elemental chlorine is connected to both the horrors of modern warfare and the establishment of a better health care for the entire world. Chlorine is certainly made in large amounts through the chlor-alkali process, but it is also made in smaller quantities through a high-temperature electrochemical process in which molten sodium chloride is electrolyzed, usually in a refractory container. This apparatus is called a Downs cell. The chemistry can be represented very simply, as shown in Scheme 7.2.

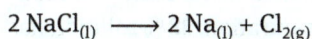

$$2\,NaCl_{(l)} \longrightarrow 2\,Na_{(l)} + Cl_{2(g)}$$

Scheme 7.2: Downs cell reaction.

This is a simple oxidation–reduction reaction which can also be classed as a decomposition reaction. What it does not show is that calcium chloride is also mixed into the melt, at up to 58 % $CaCl_2$ to 42 % NaCl, because this lowers the melting temperature of the mix. While it is used for the production of chlorine, sodium metal is the primary useful product.

Like sodium hydroxide, chlorine has many uses, some of them as intermediates to further chemicals or to user end products. Also like sodium hydroxide, several of the uses of chlorine can be viewed as cross-over items, which end up being part of what is considered an organic material. Table 7.2 shows several such uses.

7.3.3 Hydrochloric acid

Hydrochloric acid, also still called muriatic acid, can be produced in a number of ways. The direct combination of hydrogen gas and chlorine gas is usually called hydrogen chloride when this is done. Although this is a simple reaction, it is not always

Table 7.2: Use of chlorine.

Cl$_2$ Derivative	Use or end product	Comment
Bleach	NaOCl, disinfectant	
Elemental Cl$_2$	Water purification	
Halogenated solvents	Electronics, dry cleaning fluids	[8]
Organo-chlorine fine chemicals	Further fine chemical and drug or medicine synthesis	C–Cl bond can be broken selectively when compared to C–C or C–H bonds
Polyvinyl chloride	Piping, plastics applications	
Vinyl chloride	Monomer, used for production of polyvinyl chloride	

a cost-effective way to make the acid on an industrial scale. The reaction chemistry is shown in Scheme 7.3.

$$H_{2(g)} + Cl_{2(g)} \longrightarrow 2\,HCl_{(g)}$$

Scheme 7.3: Production of hydrogen chloride.

The resulting hydrogen chloride can be dissolved into water to make stable solutions of hydrochloric acid. While such solutions can vary in concentration, 38 % solution has become a standard, because it represents a good balance of ease of handling and evaporation rate. Several companies produce hydrochloric acid for a wide array of uses. ERCO Worldwide describes this as:

> "Hydrochloric acid (HCl) is used in a variety of ways including as a catalyst in synthesis and regeneration, recovery of semi-precious metals from used catalysts, pH control, regeneration of ion exchange resins used in wastewater treatment and electric utilities, neutralization of alkaline products or waste materials, and in brine acidification for use in the production of chlorine and caustic soda...Other uses of HCl include: manufacture of dyes and pigments; removal of sludge and scale from industrial equipment; deliming, tanning and dying of hides by the leather industry; manufacture of permanent wave lotion; carbonizing of wool; assisting in bleaching and dyeing in the textile industry; and purification of sand and clay [9]."

This points to the interesting fact that when organic compounds are chlorinated with elemental chlorine, the by-product is HCl. When such reactions are done on a large scale, this then results in a large amount of HCl. This product is then captured and sold as hydrochloric acid. Despite its toxic and corrosive nature, the ability to ship large amounts of hydrochloric acid is well established. Hydrochloric acid has the number 1789 on a Department of Transportation diamond used in Canada, the United States, and Mexico [10]. An example of such transport can be seen in Figure 7.3.

Figure 7.3: Hydrochloric acid transport by tanker truck.

There are many uses for hydrochloric acid, but the pickling of steel remains one of the largest volume ones. "Pickling" means the removal of iron oxides, rust, from metal surfaces, so that the surface can be further treated. The reaction chemistry can be generalized to that shown in Scheme 7.4.

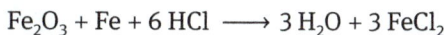

$$Fe_2O_3 + Fe + 6\ HCl \longrightarrow 3\ H_2O + 3\ FeCl_2$$

Scheme 7.4: Pickling of steel.

The final solution, sometimes called pickling liquor, is toxic enough that industrial producers have found ways to regenerate the HCl solution for later re-use.

In organic chemistry, double bonds can also react with hydrochloric acid, resulting in organo-chlorine molecules. This is another major use of hydrochloric acid. The ability of the C–Cl bond to break preferentially to the C–C or the C–H bonds in a molecule make organo-chlorines useful chemicals for further syntheses. Scheme 7.5 shows the most general version of this.

$$2\ CH_2{=}CH_2 + O_2 + 4\ HCl \longrightarrow 2\ ClCH_2CH_2Cl + 2\ H_2O$$

Scheme 7.5: Chlorination of a double bond.

7.3.4 Titanium dioxide

The production of titanium dioxide can be accomplished via two methods called the chloride process and the ilmenite process. Titanium dioxide finds numerous industrial uses, as well as consumer end uses. It depends on chlorine for the extraction of titanium from crude titanium dioxide, or rutile ore, in the chlorine process. The reaction chemistry can be written in a straightforward manner, as shown in Scheme 7.6.

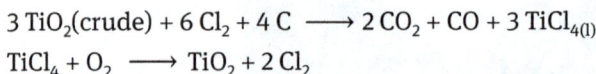

$$3\,TiO_2(crude) + 6\,Cl_2 + 4\,C \longrightarrow 2\,CO_2 + CO + 3\,TiCl_{4(l)}$$
$$TiCl_4 + O_2 \longrightarrow TiO_2 + 2\,Cl_2$$

Scheme 7.6: Titanium dioxide production.

The first reaction is routinely run at 900 °C, while the second is run at even higher temperatures, generally 1,200–1,700 °C.

The ilmenite process does not depend on chlorine, but is useful for ores with lower titanium dioxide concentrations. Scheme 7.7 shows the basic reaction chemistry.

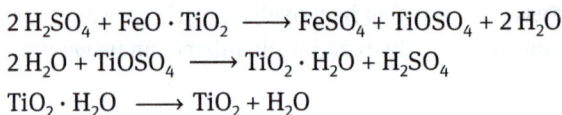

$$2\,H_2SO_4 + FeO \cdot TiO_2 \longrightarrow FeSO_4 + TiOSO_4 + 2\,H_2O$$
$$2\,H_2O + TiOSO_4 \longrightarrow TiO_2 \cdot H_2O + H_2SO_4$$
$$TiO_2 \cdot H_2O \longrightarrow TiO_2 + H_2O$$

Scheme 7.7: Ilmenite process for titanium dioxide production.

The first step in this process is the separation of iron from titanium by chemical means. After this, the sulfuric acid must be separated from the titanium fraction of the ore, so that titanium dioxide can be isolated.

The profile of uses for titanium dioxide is wide, with the major applications being paints and plastics, according to the USGS [4]. In almost all cases, titanium dioxide is considered a whitener or flattener that gives paint, paper, and other materials the proper look and reflective shine. But according to the USGS Mineral Commodity Summaries, titanium dioxide also finds uses in: "catalysts, ceramics, coated fabrics, and textiles, floor coverings, printing ink, and roofing granules", [4] at least in smaller quantities. As well, it finds use as a food additive in gums and candies.

7.3.5 Sodium chlorate

Sodium chlorate is a commodity chemical that has only one single major use: bleaching pulp in the production of paper. Yet the amounts of sodium chlorate used annually (over 100 million tons) means that it represents a major use of chlorine. Scheme 7.8 shows the reaction chemistry for the manufacture of sodium chlorate which occurs as an electrolysis in a hot solution.

$$3 H_2O + NaCl \longrightarrow NaClO_3 + 3 H_2$$

Scheme 7.8: Sodium chlorate production.

The pH of this reaction must be strictly controlled to obtain sodium chlorate as a product, and the temperature is elevated and kept at 70 °C.

As mentioned, the uses for sodium chlorate are dominated by the bleaching application. HIS, a major producer of sodium chlorate, states at its web site:

> "In 2020, this application represented over 90 % of total global consumption. Other minor uses include weed control, production of potassium chlorate and sodium chlorite, and several other smaller applications" [11].

7.3.6 Sodium silicate

Sodium silicate, also still known by the older name "water glass", is made through the reaction of sodium hydroxide and silica. Scheme 7.9 shows the reaction chemistry for what is called liquid phase production.

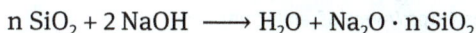

$$n SiO_2 + 2 NaOH \longrightarrow H_2O + Na_2O \cdot n SiO_2$$

Scheme 7.9: Liquid phase sodium silicate production.

The starting silica is introduced to the reaction as fine sand, where it is mixed with sodium hydroxide and water. Steam is then introduced to both raise the temperature of the reaction and to drive it to the products.

The firm IHS Chemicals produces both sodium chlorate and sodium silicate, and concerning the latter, states at its web site:

"World consumption of sodium silicate is forecast to grow by 1.8 % annually during 2021–26. Mainland China is the leading consumer, accounting for about 33 % of total world consumption, followed by Western Europe, the United States and Other Asia. In addition to a variety of direct uses, sodium silicate is consumed in the downstream production of derivative silicates, silicas and aluminosilicates, including zeolites. These derivatives account for a substantial percentage of total sodium silicate consumption" [11].

It is important to note here that sodium silicate is routinely used in "downstream production" of other commodity chemicals. Thus, end uses are somewhat difficult to find, although chemists and chemical engineers are almost always familiar with the laboratory uses of silica gel.

7.4 Recycling and re-use

There are no recycling programs for salt or the commodity chemicals made from it. Salt is consumed or chemically reacted in all of the processes discussed in this chapter, and thus there is no re-use or recycling of what are ultimately starting materials.

We have seen that hydrogen chloride can become a by-product in other industrial processes, and thus the production and sale of hydrochloric acid can be considered a form of chemical re-use.

Bibliography

[1] The Salt Institute. LinkedIn website. (Accessed 19 December 2023, as: Linkedin.com/company/salt-institute/about).
[2] EUSalt, European Salt Producers Association. Website. (Accessed 19 December 2023, as: https://pr.euractive.com/company/eusalt-european-salt-producers-association-87894#).
[3] Salt Partners. Website. (Accessed 19 December 2023, as: https://salt-partners.com).
[4] United States Geological Survey, Mineral Commodity Summaries, 2023. Website. (Accessed 18 December 2023 as: https://www.usgs.gov, https://doi.org/10.3133/mcs2023, as a downloadable pdf).
[5] US Salt. Website. (Accessed 19 December 2023, as: https://www.ussaltllc.com).
[6] Morton Salt. Website. (Accessed 19 December 2023, as: https://www.mortonsalt.com).
[7] World Chlorine Council. Website. (Accessed 19 December 2023, as: https://worldchlorine.org/).
[8] Halogenated Solvents Industry Alliance, HSIA. Website. (Accessed 19 December 2023, as: https://hsia.org).
[9] ERCO Worldwide. Website. (Accessed 19 December 2023, as: https://www.ercoworldwide.com).
[10] a. Environmentalchemistry.com. Website. (Accessed 19 December 2023, as: https://environmentalchemistry.com/yogi/hazmat/ b. Emergency Response Guidebook (ERG). Website. (Accessed 19 December 2023, as: phmsa.dot.gov/training/hazmat/erg/emergency-response-guidebook-erg).
[11] IHS Chemicals. Website. (Accessed 19 December 2023, as: https://cdn.ihs.com/www/pdf/HIS-Chemical.pdf, and as: https://cdn.ihs.com/www/pdf/Chemical-Economics-Handbook-Brochure.pdf).

8 Fluorine, fluorite, and fluorine-based materials

8.1 Introduction

Fluorine, the lightest of the halogens, was the last to be isolated as a pure element, even though fluoride minerals had been known and used for centuries. Indeed, the word "fluorine" comes from a Latin term meaning "to flow", because fluoride compounds acted as fluxes in metal refining, lowering the working temperature for medieval metallurgists when they were purifying different metals. This was well known in medieval times, and fluorite continues to be used in this capacity today. But the ferociously reactive nature of elemental fluorine is such that none exists as the free element on Earth, and it is extremely difficult to isolate. Rather, all fluorine exists as fluorides in one of a few minerals, and fluorine compounds can be formed with most other elements, including the heavier elements among the noble gases.

Fluorine is found predominantly in the minerals fluorite, fluorapatite, and cryolite. All of these can be mined profitably, if they can be located. Cryolite is rare enough in large deposits that it is no longer mined, and fluorite provides most of the needs of industry for fluorine.

Attempts to isolate and purify fluorine were undertaken in earnest in the nineteenth century, but it required several decades before anyone succeeded in the isolation. Indeed, there were several injuries and even fatalities in the numerous experiments designed to extract the element from its compounds. The honor of fluorine's "discovery" finally went to Henri Moissan, who was able to make the claim in 1886 after years of work and several unsuccessful attempts. His successful set-up involved a cooled apparatus, platinum–iridium electrodes, and other metal parts (because fluorine even corrodes platinum metal, and the alloy was found to be less reactive), and fluorspar stoppers. He was awarded the 1906 Nobel Prize in chemistry for this achievement, and sadly and ironically, passed away only a few months after the receipt of the honor. Interestingly, the technique by which he isolated this reactive element continues to be used today, with only minor changes and improvements.

Because of the extreme reactivity of fluorine, most fluorine-containing minerals are not actually refined to the point of isolating elemental fluorine. Rather, hydrofluoric acid, HF, is isolated and used in numerous applications, since it delivers the necessary fluorine in any process.

https://doi.org/10.1515/9783111329512-008

8.2 Calcium fluoride production

Calcium fluoride (CaF$_2$), also known as "fluorspar" or "fluorite", is mined in several countries, and is tracked by the USGS Mineral Commodity Summaries, in large part because of the need for it in refining metals which are strategically important [1]. Interestingly, even though the formula for this material is quite simple, it can occur naturally with numerous different impurities, which often depend on where it is mined. This is why it is sometimes found as a deeply colored material (impurities that are transition metal ions tend to impart color to the mineral), whereas when it is purified, it is a white solid. Many times, the impurities include rare earth elements or thorium, and the extraction of these elements can be the economic driving factor for mining the ore. This will be discussed in more detail in Chapter 15. Fluorite production, in thousands of metric tons, is shown in Figure 8.1. The United States is not listed in the figure, as the USGS states proprietary concerns prevent this.

It is apparent from the figure that China currently dominates the world production of fluorite, although the lack of data from the United States skews this figure somewhat. While this is connected in part to China's production of metals such as iron and steel, it is also associated with China's current production of rare earth element oxides.

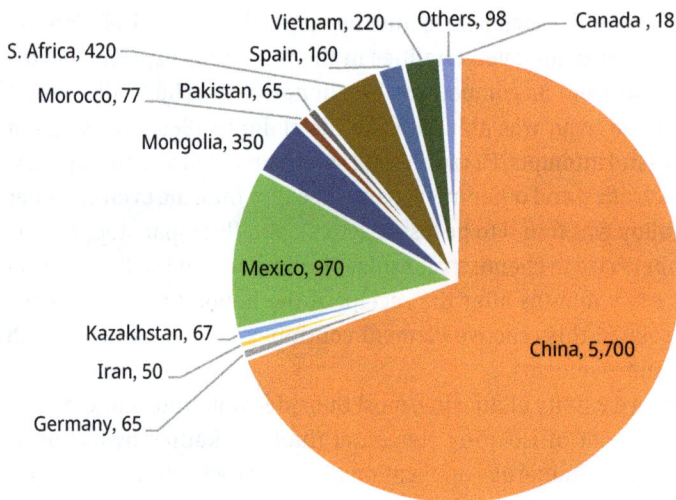

Figure 8.1: Fluorite production, in thousands of metric tons.

8.3 Fluorine isolation

When fluorine is isolated as an element, the isolation can be accomplished in an electrochemical cell utilizing both KF and HF. Scheme 8.1 shows the reaction chemistry.

$$KF + HF \longrightarrow KHF_2$$

followed by

$$2\,KHF_2 \longrightarrow F_{2(g)} + H_{2(g)} + 2\,KF$$

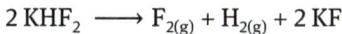

Scheme 8.1: Isolation of fluorine gas.

This reaction chemistry and the experimental apparatus has not changed appreciably from Moissan's original apparatus of the late 1800s. There has been one recent report of a chemical means by which fluorine can be prepared, but this is not used on an industrial scale.

Elemental fluorine does find some uses, mostly in the production of uranium hexafluoride and of sulfur hexafluoride. The simplified reaction chemistry is shown in Scheme 8.2.

$$U + 3\,F_{2(g)} \longrightarrow UF_{6(g)}$$

and

$$S + 3\,F_{2(g)} \longrightarrow SF_{6(g)}$$

Scheme 8.2: Production of UF_6 and of SF_6.

Both of these products have significant uses of their own. Since it is a gas, uranium hexafluoride can be isotopically enriched in U-235 through centrifugation. This isotope is the fissile form of uranium, but is a small amount of any uranium source. Most is the isotope U-238, which is inactive in terms of fissile decay. As well, since fluorine has only one isotope (F-19), differences in the UF_6 gas masses are entirely the result of the two isotopes of uranium.

Sulfur hexafluoride may not be a material with which people are familiar, at least not in terms of user end products, but it can be compressed to a liquid for transport, finds use as a dielectric medium, and thus is made each year on a scale of almost ten thousand tons. To a smaller extent, it is also used as an insulator in windows.

Uranium hexafluoride is generally made through a reaction with uranium dioxide and hydrofluoric acid, as seen in Scheme 8.3.

Several thousand tons of fluorine are used each year in this process. This isotopic enrichment is required to obtain both fuel for nuclear power plants and material for

$$UO_2 + 4\,HF \longrightarrow UF_4 + 2\,H_2O$$

followed by

$$UF_4 + F_{2(g)} \longrightarrow UF_{6(g)}$$

Scheme 8.3: Production of UF_6 with HF.

weaponry. Thus, operations in which such enrichment occurs are carefully monitored by national governments.

8.4 Metspar and acidspar

The two terms "metspar" and "acidspar" refer to different grades of fluorite, the first being 60–85 % fluorite and the second being the higher purity, 97 %, fluorite.

Metspar is used predominantly as part of the process of smelting iron. Usually, 3–4 kg of metspar are added to each batch of molten metal as steel is being refined. This lowers both the working temperature for the batch, and lowers the viscosity of the molten material, making it easier to pour and use.

Acidspar is of high enough purity to be used in the production of hydrofluoric acid, HF, the starting material for numerous fluorine-containing materials and user end products.

For the past several years, over 3½ million tons of fluorite have been mined annually, and separated into these two grades of material.

8.5 Hydrofluoric acid

In many cases, in which fluorine is an important element or component of a chemical process, elemental fluorine is not required to produce the target material. Rather, hydrofluoric acid, HF, is used as a starting material. To produce it, acidspar is reacted with sulfuric acid to form the hydrofluoric acid, according to the reaction shown in Scheme 8.4.

$$CaF_2 + H_2SO_4 \longrightarrow 2\,HF + CaSO_4$$

Scheme 8.4: Hydrofluoric acid production.

This then becomes another industrial use for sulfuric acid, discussed in Chapter 2, and thus another commodity that ultimately affects the price of sulfuric acid.

$$3\ NaOH + 6\ HF + Al(OH)_3 \longrightarrow Na_3AlF_6 + 6\ H_2O$$

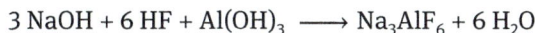

Scheme 8.5: Synthetic cryolite production.

The production of synthetic cryolite for the further production of aluminum metal is one significant use of fluorine in the form of HF. The reaction chemistry is shown in Scheme 8.5.

In the past, cryolite has been mined, but since it is a much less common mineral than fluorite, it is now economically more feasible to produce the cryolite to be used in aluminum smelting. The last operating cryolite mine, in Greenland, was closed in 1987 because it was no longer economically profitable to continue the operation.

Since aluminum refining is a major industrial process with aluminum finding a multitude of end uses, the need for cryolite will continue as long as there is a demand for aluminum. Aluminum production will be discussed in more detail in Chapter 12 [2].

8.6 Teflon

Teflon, more properly known as polytetrafluoroethylene (PTFE), is one of the major advances in materials chemistry of the twentieth century. People now find it difficult to imagine life without such things as Teflon pans and cookware, as well as Gortex®clothing for outdoor wear in rough weather. The story of Teflon's discovery is one of both serendipity and thorough, well-executed science, originally with Roy J. Plunkett at DuPont in the 1930s, and later at W.L. Gore in the 1960s. While the DuPont research labs had discovered and characterized Teflon, it is generally believed that the Manhattan Project, the United States' effort at building an atomic weapon during the Second World War, was the driver for scaling up Teflon production to an industrial-scale level. The connection is that some inert coating was needed for the piping through which isotopic enrichment of U-235 in the form of UF_6 gas was being undertaken. It turned out that Teflon coating was less reactive, and thus more useful, even than nickel metal had been. Nickel had been used in large amounts prior to its replacement with the PTFE coating.

Since those early efforts, DuPont has recognized the value of continuing to expand the potential uses of Teflon, and in 2014 celebrated 75 years of its DuPont Plunkett Awards, which was stated in a press release at the time as, "given in recognition of new, cutting-edge applications for DuPont fluoropolymers" [3–5].

The production of tetrafluoroethylene and the subsequent production of PTFE are examples of how a process can straddle both organic and inorganic chemistry. Scheme 8.6 shows how tetrafluoroethylene is made, and Scheme 8.7 shows the basic olefin-based polymerization to make Teflon. The latter is generally a free radical polymerization.

$$2\,HF + CHCl_3 \longrightarrow CHClF_2 + 2\,HCl$$

followed by

$$2\,CHClF_2 \longrightarrow F_2C{=}CF_2 + 2\,HCl$$

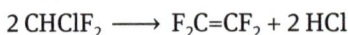

Scheme 8.6: Tetrafluoroethylene production.

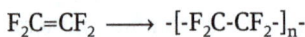

$$F_2C{=}CF_2 \longrightarrow \text{-}[\text{-}F_2C\text{-}CF_2\text{-}]_n\text{-}$$

Scheme 8.7: Teflon production.

Chloroform is the starting material for this monomer's production, and thus the production of chloroform becomes a major component of the production of Teflon, while HF is the other major feed to produce tetrafluoroethylene. Going back further, chloroform is made from methane and chlorine, which reinforces the idea of a cross-over or merging of organic and inorganic chemistry to create this material. As well, the second reaction requires a substantial input of energy, and is usually run at 550 °C or higher.

Once again, this appears to be a reaction that can be categorized one within the domain of organic chemistry, even though strict definitions of the field tend to insist on the presence of some form of carbon–hydrogen covalent bond (which is missing here).

Because the solubility of PTFE is low in all solvents, the polymerization reaction is conducted in water, but as an emulsion. The resulting small particles of PTFE are then formed into whatever shape is required for the end use.

8.7 Perfluorooctanoic acid

Perfluorooctanoic acid (PFOA) is a material that has been produced on an industrial scale for the past seven decades. It finds numerous uses in different end products, but sees a large use in the polymerization production of Teflon.

The reaction chemistry for the production of PFOA is somewhat difficult to write as a simple reaction, since different co-products form and their percentages can vary with differences in reaction conditions. The process is referred to as an electrochemical fluorination, and usually requires 5–6 V to produce products. Scheme 8.8 shows the basic reaction chemistry, in a simplified form.

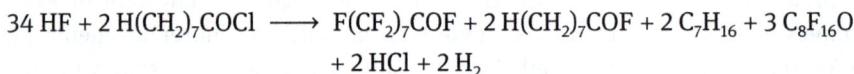

$$34\,HF + 2\,H(CH_2)_7COCl \longrightarrow F(CF_2)_7COF + 2\,H(CH_2)_7COF + 2\,C_7H_{16} + 3\,C_8F_{16}O$$
$$+ 2\,HCl + 2\,H_2$$

Scheme 8.8: PFOA production.

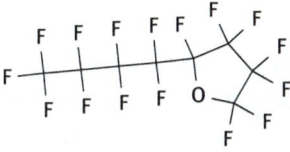

Figure 8.2: Lewis structure of FC-75.

The desired product sometimes forms in as low as 10 %, but the by-products are also useful. For example, the product listed as $C_8F_{16}O$ can be a perfluorinated cyclic ether with the structure shown in Figure 8.2. It is an isomer of the perfluorooctanoic acid fluoride which itself must be hydrolyzed to produce PFOA. Because of the insolubility of this ether in water, its ability to function as an inert coolant, and the large amount of it produced as a secondary product in this reaction, 3M markets this as FC-75, one of its Fluoinert fluids. Indeed, 3M states of these materials: "A family of perfluorinated liquids offering unique properties ideally suited to the demanding requirements of electronics manufacturing, heat transfer and other specialized applications" [6].

PFOA has proven to be a remarkably persistent chemical, and has been found in ultra-trace amounts in widely dispersed areas of the world, and has also been found in tiny amounts in human blood serum during different studies. Currently, there is no agreement as to whether or not the presence of trace amounts of PFOA in the environment is detrimental to human and other life.

8.8 Fluorine-containing fibers

Less than 30 years after the discovery, characterization, and scale-up of Teflon to an industrial scale process, Teflon fibers were pioneered by W.L. Gore and Associates. Once again, a certain amount of serendipity were evident in the discovery of Goretex, as the slow extrusion of PTFE fibers was found to be too time consuming, and a sudden jerk on the heated material did not break the material, but rather produced fibers that were largely air by volume.

8.9 Teflon and fluorine-containing fiber uses

It is perhaps obvious that the average consumer considers frying pans and other cookware to be the major consumer end use of fluorinated materials like Teflon, and waterproof clothing to be the main use of fluorine-containing fibers. It will then come as something of a surprise to realize that the largest use for PTFE is in cables and wiring, because it is an excellent insulator.

8.10 Recycling and re-use

Recycling of fluorine-containing materials is not a matter of recycling consumer end products (like recycling aluminum, plastic, glass, or paper), but rather a matter of ensuring that materials such as hydrofluoric acid or by-product hydrochloric acid are re-used and not released into the atmosphere. The driving force for this is at least indirectly economic, as companies face stiff fines for the accidental release of harmful materials into their local environments.

Bibliography

[1] United States Geological Survey, Mineral Commodity Summaries, 2023. Website. (Accessed 18 December 2023 as: https://www.usgs.gov, https://doi.org/10.3133/mcs2023, as a downloadable pdf).
[2] Alcoa. Website. (Accessed 20 December 2023, as: https://www.alcoa.com).
[3] DuPont. Website. (Accessed 20 December 2023, as: https://www.dupont.com).
[4] The Plunkett Awards. Chemours. Website. (Accessed 20 December 2023, https://www.chemours.com/en/about-chemours/our-businesses/advanced-performance-materials/plunkett-awards).
[5] W.L. Gore, Inc. Website. (Accessed 20 December 2023, as: https://www.gore.com/products/categories/fabrics?view=gore-tex-fabrics).
[6] 3M, Fluorinert. Website. (Accessed 20 December 2023, as: https://www.3m.com/en_US/p/d/b40045180/).

9 Borderline inorganics–organics

9.1 Introduction

There are some commodity chemicals that can be considered inorganic in their major uses and applications, but organic in their sourcing, as well as some for which the reverse situation is true. Those we discuss here may not be a complete list, since different people will define these materials in different ways, but do represent several examples of industrial-scale commodities that in some way straddle the line between organic and inorganic chemistry.

9.2 Carbon black (or: channel black, colloidal black, furnace black, and thermal black)

Carbon black is made by the combustion of hydrocarbons in a lean oxygen environment, and is manufactured on the level of millions of tons each year. Because it is often produced from the heavier fractions of petroleum, through incomplete combustion, it is often considered an organic material. But since there is no bonding of carbon to any other element, it is often considered an inorganic material. Scheme 9.1 shows an idealized chemistry for the synthesis of carbon black.

$$H\text{-}[CH_2]_n\text{-}H \longrightarrow n+1\ H_{2(g)} + C_{(s)}$$

Scheme 9.1: Carbon black synthesis.

9.2.1 Carbon black production

The production of carbon black is a mature industry, with several producers internationally. A listing of companies that produce this material includes the following (listed alphabetically, and not by quantities produced):
– Alexandria Carbon Company – operates in Egypt under the parent firm Aditya Birla.
– Cabot Corporation – sells over 2 million tons of carbon black annually, under several different trade names.
– Cancarb Limited – headquartered in Alberta, Canada, producing what the company calls "thermal carbon black."
– Columbian Chemicals – another firm under the Aditya Birla, based in Georgia, USA. Produces carbon black under several different trade names.
– Continental Carbon, which is owned by China Synthetic Rubber Corporation.

https://doi.org/10.1515/9783111329512-009

–   Evonik – now operating as Orion Engineered Carbons, since its sale in 2011 to Rhone Capital and Triton Partners.
–   Sid Richardson – purchased by Tokai Carbon, Ltd., headquartered in Japan, where they sell more than 30 grades of carbon black.
–   Timcal Graphite and Carbon – produces carbon black as one product among many in its suite of carbon-based materials.

These companies produce carbon black through what is called the furnace black process or the thermal black process [1–6]. The furnace black process has been the most common for the past several decades, and involves blowing the feedstock into a stream of high temperature gas to effect the partial combustion. This method is well suited for making carbon black of various particle sizes, depending on how conditions are adjusted. The thermal black process is used by Cancarb, and involves injecting natural gas into pre-heated chambers, and cooling the product with water. The carbon black is then collected in bag filters, and the hydrogen is used to produce heat to continue the process [3].

9.2.2 Carbon black uses

Many people, including chemists, have no real idea what carbon black is used for, despite its enormous annual production. While there are many uses for the material, one dominates all the others – the addition of carbon black to tires. This is not simply because consumers expect tires to look black. Rather, the carbon black acts to strengthen the rubber, and produces more durable tires than if it was not added. At times, up to half the mass of special tires can be carbon black. Other uses include:
–   Printer toner and inks
–   Radar absorbent coatings
–   Other-than-tire rubber materials (such as hosing and belts)
–   Pigments
–   Filler in paints and coatings

Additionally, carbon black is used in various niche markets.

9.2.3 Carbon black, recycling, and re-use

Because carbon black is always mixed into some other product, often a consumer end-use product, any recycling is the recycling of the end product. For example, tires are recycled in many parts of the world. Tires can be re-used, but when their usable life as a tire is complete, they have been used in creating reinforcing walls for road embankments, and are sometimes shredded for use as a material for playgrounds.

9.3 Sodium tri-poly-phosphate

Sodium tri-poly-phosphate, often known and sold as STPP, can be considered another material made from the electrochemical reaction of a brine solution, but can also be considered a material that starts with phosphate. The USGS Material Commodity Summaries does not track STPP directly, but does track phosphate rock, and its production [7]. Within the United States, most phosphate is mined in Florida and North Carolina, although there are other operations in western states as well. Worldwide production was shown in Figure 5.2, in the discussion of fertilizers.

9.3.1 STPP production

Broadly, the reaction of phosphoric acid and sodium carbonate is how STPP is formed. Chemically, it is simply an addition, as shown in Scheme 9.2.

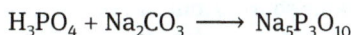

$$H_3PO_4 + Na_2CO_3 \longrightarrow Na_5P_3O_{10}$$

Scheme 9.2: Production of sodium tri-poly-phosphate (STPP).

More specifically, the addition of monosodium phosphate and disodium phosphate with controlled heating is how STPP is produced on an industrial scale. The reaction chemistry can be represented as that shown in Scheme 9.3.

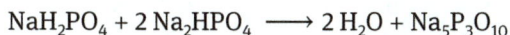

$$NaH_2PO_4 + 2\,Na_2HPO_4 \longrightarrow 2\,H_2O + Na_5P_3O_{10}$$

Scheme 9.3: Production of STPP with di- and mono-sodium phosphate.

When food grade STPP is the end result, the chemistry is somewhat more complex. The USGS Mineral Commodity Summaries states: "About 25 % of the wet-process phosphoric acid produced was exported in the form of upgraded granular diammonium phosphate (DAP) and monoammonium phosphate (MAP) fertilizer and merchant-grade phosphoric acid. The balance of the phosphate rock mined was for the manufacture of elemental phosphorus, which was used to produce phosphorus compounds for industrial applications, primarily glyphosate herbicide" [7]. The reason of this level of refining is the need to remove any naturally occurring impurities before the final STPP materials are used in foods. Major producers of STPP include the following, listed in alphabetical order:
- Biddle Sawyer Corporation – headquartered in New York City.
- Hubbard Hall – in Connecticut, USA

–   Ortho Chemicals Australia Pty, Ltd. – in Victoria, Australia, sells a wide product line of bulk chemicals.
–   Prayon Deutschland, GmbH – based in Dortmund, advertises itself as a "worldwide leader in phosphate creativity."
–   Thatcher Company – in Utah, USA also has a very wide product line, including numerous materials that are not phosphates.
–   Tianjin Ronghuiyuanyang International Trade Company, Ltd. – in Tianjin City, China.

Collectively, several million tons of STPP are produced annually, in several different grades [8–13].

9.3.2 STPP uses

STPP is placed here with the other materials that can be considered a cross-over from organic to inorganic because it starts with purely inorganic sources, but can be used as a food additive among other things. It is a food preservative for meats and sea foods; in Europe it has the number E451, and according to the United States Food and Drug Administration, it is a food additive that is generally recognized as safe (GRAS). In general, its function in foods is as an additive that retains moisture.

STPP is also used in the following applications, with some of them in very large amounts:
–   Laundry and dish detergents
–   Toilet cleaner
–   Water treatment
–   Toothpaste additive
–   To help disperse pigments
–   Paper coating, to resist oils
–   Dispersant in ceramics
–   Leather tanning

The use of STPP as a detergent is large enough that entire trade organizations exist devoted to detergent manufacture and use. The American Cleaning Institute [14], the UK Cleaning Products Industry Association [15], and the Japan Soap and Detergent Association [16] are three organizations with interests in STPP production, although there are others as well [17].

9.3.3 STPP recycling and re-use

Since all STPP uses are for the production of other materials, or are in consumer end-use products, there is no recycling of it. There is however a continuing interest in determining how STPP as a detergent can be controlled and dealt with when it makes its way into effluent water streams, since excess cleaning materials in water can cause serious environmental problems.

9.4 Borax and borates

The mineral borax, also called sodium tetraborate, is routinely mined as a decahydrate. Thus, the material has the formula $Na_2B_4O_7 \cdot 10\,H_2O$. But borate minerals include kernite or rasorite ($Na_2B_4O_6(OH)_2 \cdot 3\,H_2O$), tincal (essentially another term for borax), and ulexite ($NaCaB_5O_6(OH)_6 \cdot 5\,H_2O$), all of which can be mined if there is economic incentive. It has been known for decades that borax can be used as a detergent, and indeed, at least one borax-containing laundry detergent is still marketed [18].

The American Borate Company states on its web site that it is the largest producer of boron chemicals in North America, claiming:

"Borates impact our everyday lives. They are critical to the growth of healthy plants, used in metallurgy, found inside fiberglass and specialty glass, incorporated into gypsum to construct walls and ceilings and are a component of detergents…" [19]

It is the use of boron compounds in agriculture that represents a crossing point from inorganic to organic chemistry.

9.4.1 Borax production

The USGS Mineral Commodity Summaries do track the production of such materials globally, under the generic heading, "Boron." In the United States, borax is mined by two firms in southern California. Figure 9.1 shows the production breakdown by country, with the US production withheld (the USGS cites proprietary data). Purification usually involves crystallization from a hot, aqueous solution, often at elevated pressure. At least one patent details how this process removes insoluble silicate impurities, and how aluminum-containing salts are added to remove soluble silicate impurities [20].

The relatively small number of countries that produce borax is an indicator that borax-containing ores are unevenly distributed throughout the Earth's crust. Significant deposits occur in those countries that are listed here.

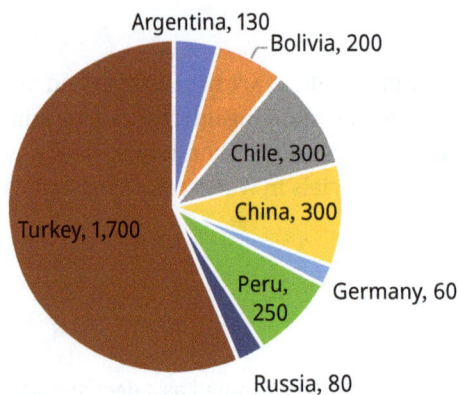

Figure 9.1: International boron production (in thousands of metric tons B_2O_3).

9.4.2 Borax uses

There are several different industrial needs for borate compounds, but the USGS notes that, "Although borates were used in more than 300 applications, more than three-quarters of the world consumption was used in ceramics, detergents, fertilizers, and glass." [7]

A great deal of boron chemistry, ultimately starting from borax, was undertaken in the 1950s, sponsored in large part by the United States Air Force. The aim was to produce large quantities of boranes (a series of boron hydride compounds) to use as jet and rocket fuel, since these would enable faster flights with fuels that weigh less than comparable hydrocarbons. The chemistry behind this is very rich, and has expanded to include carbon–boron cluster chemistry, all of which has been encompassed in at least one book [21]. However, the large-scale production of boranes ceased in the 1960s, because they proved to be too difficult to handle and too corrosive to the jet engine components when compared to more traditional hydrocarbons [22].

9.4.3 Borax re-use and recycling

Borax and any boron-containing compounds derived from it are routinely consumed, or end up in some consumer end product. Thus, there are no recycling programs for borax.

9.5 Asphalt

This widely used road material is a hybrid of inorganics – meaning crushed stone – and the heaviest fraction left over from crude oil refining. It is sometimes still called "bitumen" or "tar", and does have uses other than road materials.

9.5.1 Asphalt source materials, formulas, and production

The production of crude oil into motor fuel and monomers for polymerization is obviously a massive one that rearranges molecules into those of desired molecular weight, often by breaking larger molecules into smaller ones. What is called the C8 fraction is the target for the isomers that become motor fuels. But this process leaves behind heavy, viscous, and difficult-to-process materials which are often collectively called tar. This material is the organic component for asphalt. The inorganic component is crushed stone. In the past 30 years, several other additives have been developed so that asphalt can be used under extreme weather conditions. It is difficult to represent asphalt production using any sort of reaction chemistry simply because all asphalt mixtures are just that mixtures. There is never anything resembling a stoichiometric mixture of the components in asphalt.

When this aggregate is used as a road-paving material, it is more correctly called asphalt concrete. Such mixtures are routinely as high as 90 % stone and other inorganic material. Some lighter hydrocarbons, such as kerosene, do get added to asphalt before it is put in place, so that the batch is less viscous and easier to work.

Enough asphalt is used in the developed and developing world that there are trade organizations devoted to it [23–28].

9.5.2 Asphalt uses

While many people consider paving and roads to be the only use of asphalt, and indeed this one use accounts for almost 90 % of asphalt use, there are some others as well. Roofing can be finished and sealed with asphalt. This is often done for large, flat rooves in corporate or industrial buildings, although some residential homes do use asphalt in this way. In addition, asphalt roofing shingles are made from an aggregate that is smaller and finer, but that when mixed with the organic component adheres to the shingle base. Such shingles are used because they repel water very well.

9.5.3 Asphalt re-use and recycling

The term recycling generlly implies consumer commodities such as paper, aluminum, or glass, not materials such as asphalt. Yet almost all asphalt can be re-used, for instance, when a road is being re-surfaced. While people do not generally consider asphalt a recyclable material, virtually none is removed from a road site during re-surfacing or road work and discarded. The aggregate is routinely mixed with more hydrocarbon material and re-placed at the construction site.

Bibliography

[1] Aditya Birla Carbon Black. Website. (Accessed 20 December 2023, https://www.birlacarbon.com).
[2] Cabot. Website. (Accessed 20 December 2023, as: https://www.cabotcorp.com/solutions/products-plus/carbon-blacks-for-elastomer-reinforcement).
[3] Cancarb Limited. Website. (Accessed 20 December 2023, as: https://cancarb.com).
[4] Continental Carbon Company. Website. (Accessed 20 December 2023, as: http://www.continentalcarbon.com/carbon-black-products.asp).
[5] Tokai Carbon Co., Ltd. Website. (Accessed 20 December 2023, as: https://www.tokaicarbon.co.jp/en/).
[6] Imerys Graphite & Carbon. Website. (Accessed 20 December 2023, as: https://www.imerys.com).
[7] United States Geological Survey, Mineral Commodity Summaries, 2023. Website. (Accessed 18 December 2023 as: https://www.usgs.gov, https://doi.org/10.3133/mcs2023, as a downloadable pdf).
[8] Biddle Sawyer Corporation. Website. (Accessed 20 December 2023, as: http://www.biddlesawyer.com).
[9] Hubbard Hall. Website. (Accessed 20 December 2023, as: https://www.hubbardhall.com).
[10] Ortho Chemicals Australia Pty, Ltd. Website. (Accessed 20 December 2023, as: https://orthochemicals.com).
[11] Prayon. Website. (Accessed 20 December 2023, as: https://www.prayon.com).
[12] Thatcher Group. Website. (Accessed 20 December 2023, as: https://tchem.com).
[13] Tianjin Rong Hui Yuan Yang International Trade Company, Ltd. Website. (Accessed 20 December 2023, as: http://www.66357.tradebig.com).
[14] American Cleaning Institute. Website. (Accessed 20 December 2023, as: https://www.cleaninginstitute.org).
[15] UK Cleaning Products Industry Association. UKCPI. Website. (Accessed 20 December 2023, as: https://www.ukcpi.org).
[16] JSDA, Japan Soap and Detergent Association. (Accessed 20 December 2023, as: https://jsda.org/e/index.html).
[17] AISE, International Association for Soaps, Detergents, and Maintenance Products. Website. (Accessed 20 December 2023, as: https://www.aise.eu).
[18] 20 Mule Team Borax. Website. (Accessed 20 December 2023 as: https://www.20muleteamlaundry.com).
[19] American Borate Company. Website. (Accessed 20 December 2023, as: http://www.americanborate.com).
[20] Method of Purifying Borax. US 1812347 A. Website. (Accessed 20 December 2023, as: http://www.google.com/patents/US1812347).
[21] Grimes, R.N. Carboranes, 2nd Edn., 2011, ISBN-13: 978-0123741707.
[22] Andrew Dequasie. The Green Flame: Surviving Government Secrecy, 1991, ISBN: 978-0-8412-1857-4.
[23] National Asphalt Pavement Association (NAPA). Website. (Accessed 20 December 2023, as: https://www.asphaltpavement.org).
[24] International Society for Asphalt Pavements. Website. (Accessed 20 December 2023, as: https://asphalt.org).
[25] The Asphalt Institute. Website. (Accessed 20 December 2023, as: https://www.asphaltinstitute.org).
[26] Association of Modified Asphalt Producers. Website. (Accessed 20 December 2023, as: https://www.modifiedasphalt.org).
[27] European Asphalt Pavement Association. Website. (Accessed 20 December 2023 as: https://eapa.org).
[28] AAPA, Australian Asphalt & Pavement Association. Website. (Accessed 20 December 2023, as: https://www.pavetrend.com.au).

10 Water

10.1 Introduction and sources

Water may be the single most important material for life on Earth. While some might argue that air has that primacy of place, it can be said that a variety of life in water, and the generally accepted idea that life began in it, means that water remains the most important substance. Clearly, when it comes to industrial processes, a great many of them depend on water, either as the solvent or as one of the reactants.

The frustrating aspect of the water on Earth, at least from the point of view of one wishing to use it in an industrial-scale process, is that just over 97.5 % of it is saline. All freshwater, including nonsaline inland seas, glaciers, lakes, rivers, streams, clouds, and groundwater account for just under 2.5 % of the world's total water. This in turn means that most of the world's water has approximately 3.5 % sodium chloride in it, the average amount of salt in the oceans' waters. There are also significant amounts of magnesium, sulfur, calcium, potassium, and bromine in the oceans. There have even been attempts to extract valuable materials from seawater, such as gold and uranium, even though both are only trace elements in the oceans. Overall, this means that fresh water supplies can be extremely limited [1].

10.2 Purification

Most water needs to be purified before use, although the degree of necessary purification varies depending on the water source. Water that is to be used for human consumption must be free of organic pollutants as well as most salts, but that needed for an industrial process may need to be significantly cleaner. Because water is vital in so many processes, a wide array of purification techniques has been developed, and several organizations exist that are in some way tied to ensuring supplies of clean water [2–9]. Broadly, desalination and sewage treatment are the two most common means of purifying water. We will discuss each in some detail.

10.2.1 Desalination

There are numerous ways to desalinate water, and many can be constructed to industrial-scale size if needed. For example, what are called reverse osmosis water purification units (sometimes called ROWPU) can be as small as a hand-held item. These are often used in boating emergency kits. But ROWPU can be as large as the unit located off the coast of Israel, which produces fresh water from the Mediterranean Sea.

https://doi.org/10.1515/9783111329512-010

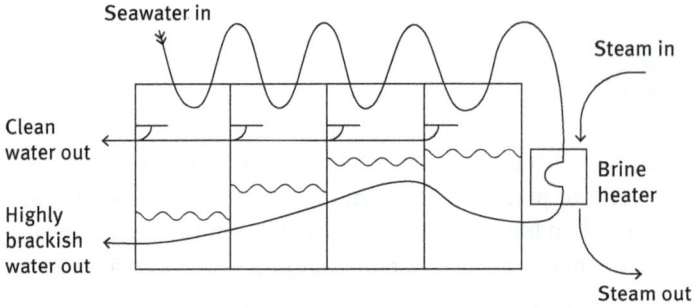

Figure 10.1: Multi-stage flash distillation operation.

There are other large desalination units off the coasts of Singapore, Saudi Arabia, and Venezuela.

While there are several different methods of desalination, one method in particular – multistage flash distillation – is used in approximately three-fourths of all operations.

Multistage flash distillation

The process by which fresh water is extracted from saline water in a multistage flash distillation operation is less a chemical process, and more a physical one, since it involves a series of heat exchangers making small portions of a brackish feed water "flash", or turn to steam, in each compartment of the apparatus. This is then re-condensed and captured as fresh water. The greatest continued expense in the life of this type of operation is the energy required to cause the flash of brackish water. A diagram of how such an apparatus works is shown in Figure 10.1.

It is noteworthy that such an operation does not extract salt from the water. Rather, water that has been flashed from the inlet brine is captured, and a more brackish solution is returned to the body of water from which it came.

Reverse osmosis

Reverse osmosis is another method that can purify large amounts of water, but in this case depends on semipermeable membranes through which water can be forced, through which dissolved particles are unable to pass. Feed water that enters the system is pressurized and exposed to the membrane. That which does not pass through is routinely passed through a pressure exchanger, where a more concentrated brackish solution is discharged, and where pressurized water can be again fed into the purification loop. Figure 10.2 shows the basic design for this.

Once again, the major long-term cost of such an operation is the energy required to maintain the pressure. Additionally, membranes do wear out and have to be replaced.

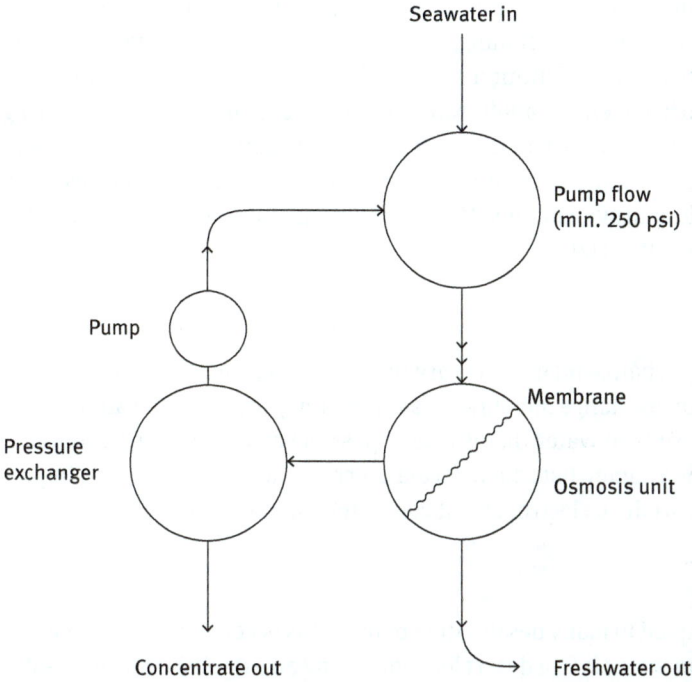

Figure 10.2: Reverse osmosis water purification.

Membrane distillation

This technique involves polytetrafluoroethylene membranes, and a thermal gradient to force water across the membrane, while impurities are removed. The first company to bring this technique to an industrial scale is Aquaver [10]. They have recently installed a plant using this technique in the Maldives, in which the heat source is the waste heat from diesel generators which produce electricity for the islands.

Electrodialysis reversal

As the name implies, this technique involves an electrical form of dialysis, particle removal from the feed water. Broadly, this can still be considered a form of reverse osmosis, although it does not depend on pressure to effect the desalination. Several companies have some interest in this technique, but GE is one company that has made significant progress to produce these water purification units, and stresses at their web site the ruggedness of the systems, and the durability of their exchange membranes [11].

Nanofiltration

This is a variation of reverse osmosis, but one that is more specifically applied to drinking water. A company that has used this technique extensively, Lenntech, describes the process at their web site: "…nanofiltration is mainly applied in drinking water purification process steps, such as water softening, decolouring and micro pollutant removal. During industrial processes nanofiltration is applied for the removal of specific components, such as colouring agents. Nanofiltration is a pressure related process, during which separation takes place, based on molecule size. Membranes bring about the separation. The technique is mainly applied for the removal of organic substances, such as micro pollutants and multivalent ions. Nanofiltration membranes have a moderate retention for univalent salts." [12]

Ion exchange

In this process, cation exchange membranes are used to replace metal cations with hydronium ions, and anion exchange membranes are necessary to replace all anions with hydroxide ions. This results in water that is free of possible cation and anion contaminants. At present, this is a somewhat smaller scale process, but has become very important in operations that produce electronic and computer components.

Solar desalination

Solar power can be coupled to many desalination plants. This becomes very useful when the area requiring clean water is in a desert location. Using an established form of solar power to drive another form of desalination operation is called indirect solar desalination. What is called direct solar involves using solar power to condense water from the nearby environment. This has traditionally been used in smaller applications, often in survival situations.

Freezing desalination

This technique takes advantages of the differences in density between ice and liquid water. The company CryoDesalination comments at their web site: "Cryo-desalination, the separation of water and salt upon freezing, utilizes the natural tendency of water to push out salt upon freezing. In practice, one utilizes energy to cool water and form ice. As ice is forming it expels most of the salt, resulting to the so called brine, which is very highly concentrated salt water. Then the ice and brine are separated followed by the warming up of the remaining ice in order to obtain the fresh water." [13]

While there are many companies that have become involved in water purification and desalination techniques, Dow Chemical has spent a significant amount of energy in producing different water treatment technologies. Indeed, their web site states: "Dow is an innovator in water purification and separation technologies, known for a number of industry firsts – including the world's first spiral-wound membrane technology for water treatment. We set the industry standard for quality and reliability with our complete portfolio of Ultra filtration (UF) and Reverse Osmosis (RO) Membranes, Fine Particle Filtration (FPF) and Electrodeionization (EDI) products. In addition, we offer the widest

range of Ion Exchange Resins that meet separation requirements from softening to ul-
trapure water generation to trace contaminant removal." [14]. Such comments are by
definition biased towards the company and sale of their products, but it is evident that
Dow has done considerable work in developing a broad range of capabilities within this
field.

10.2.2 Sewage treatment

The treatment and clean-up of fresh water that has been contaminated with some form
of domestic sewage is actually less expensive a process than the desalination methods
we have discussed. In general, municipal sewage treatment plants are smaller than the
largest desalination plants. But even the largest ones use less energy-intensive methods
of separating and concentrating components from the feed water.

Sewage treatment plants in different areas are built to somewhat different specifications depending on the expected input, but all have the following steps, or most of the
following steps, in their water cleaning process.

Pre-treatment
This routinely refers to the removal of macroscopic materials. Waste water can come
from homes and businesses, but also from storm run-off. So this step can involve the
removal of trash and yard waste as well as pieces of material that are large enough to
see but small enough that only a sieve or strainer can remove them, such as grains of
sand or dust.

Primary treatment
This step almost always involves some time period in which materials can settle from
the feed water, so that water can be separated from these materials based on density.
The reverse is also the case, meaning fats and oils that are less dense than water can be
allowed to rise to the top of the batch, for later skimming and removal. If the concentration of fats and oils is low enough, air may be blown into the batch to create a frothy
upper layer. This can then be manually skimmed and separated from the denser water.

Secondary treatment
This step very often involves some form of microbial exposure to the water that has been
cleaned through primary treatment. In many cases, a period of time elapses wherein the
microbes are allowed to consume organic material still suspended or dissolved in the
water. Then the holding tanks or bins are drained, usually through a sand bed, so that the
resulting algae or other growing material can die, dry out, and be collected for disposal.

Sludge digestion

Water that has been cleaned to this degree may leave sludge behind. The clean water may be discharged into the environment, or may be chlorinated if it is intended for subsequent human or animal use. The sludge can in some cases be sold as a fertilizer, depending upon its composition.

These four broad steps show that sewage treatment is less a series of chemical reactions, and more a series of steps based on physical differences of the components in the water. Traditionally, the purpose of such water clean-up was to prevent the spread of diseases, especially in urban areas. Only in the past few decades has this been considered a means of providing fresh water for further uses.

10.2.3 Water conservation

By far the least expensive way to increase the availability of fresh water is to conserve that which already exists. Any point that leaks in a water piping system, from the main reservoir to a residential tap, represents a loss and waste of water. Fixing leaks is almost always just a mechanical process, and the tools and parts required to do so are always well established, and far less expensive than a desalination or treatment process. To prove this point and urge people to fix this problem, the United States Geologic Survey even maintains a web site that calculates the amount of water wasted per year when a person inputs the number of dripping faucets and the drip rate. For example, a single dripping faucet, with a rate of one drip per minute, wastes 34 gallons of water per year [15].

10.3 High purity, uses

The large-scale production of electronics components and computers has seen the rise of a need for extremely high purity water. These processes require water that has no residual ions, and thus the exchange membrane processes we have discussed become requirements for the production of high purity water.

One additional method of producing extremely high purity water is the combination of elemental oxygen and elemental hydrogen to produce high purity steam which is then condensed. Such water is even free of residual oxygen if the combination is carefully controlled. Thus far, no company has brought this process up to an industrial scale, in part because other processes meet the needs of industry, and in part because of the risk of explosion in dealing with these two elements.

10.4 Uses, residential

Our modern, developed world is marked by an extensive water piping system, among other systems, such as electrical grids and highway networks. Some people in parts of the developing world still lack immediate access to fresh water, but even the governments of those poorer nations work to ensure that their people have access to fresh water, as it has become one of the markers of modern civilization.

10.5 Uses, industrial

We have already discussed several large-scale processes that require water. The following listing shows subjects in this book in which water is required as a reactant in an industrial process:
- Sulfuric acid production
- Phosphoric acid production
- Sodium hydroxymethylsulfinate production
- Chlor-alkali process
- Ostwald process
- Solvay process
- Ilmenite process
- Gold cyanide refining

Additionally, numerous processes require water at some stage, even though it is not a direct reactant. Within this book, they include:
- Frasch process
- Carbon black production
- Borax refining
- Metal refining (floatation of froths)
- The Washoe process
- Zeolite production

One other, enormous use of water is fracking and oil production, in which water is required to extract the starting material from its natural environment. These and other processes all would be impossible without large amounts of clean water.

10.6 Recycling

The section discussing sewage treatment is actually a form of water recycling. But in general, water conservation and re-use have been an important part of the developed world for decades, and is becoming more so as more people depend on sources of clean water.

Bibliography

[1]   National Ocean Service. Website. (Accessed 20 December 2023, as: https://oceanservice.noaa.gov/facts).

[2]   Aqua Europe. Website. (Accessed 20 December 2023, as: https://aqua-europe.eu).

[3]   American Water Works Association. Web site. (Accessed 20 December 2023, as: https://www.awwa.org).

[4]   American Water Resources Association. Website. (Accessed 20 December 2023, as: https://www.awra.org).

[5]   Water Quality Association. Web site. (accessed 20 December, 2023, as https://wqa.org).

[6]   British Water. Web site, (Accessed 20 December, 2023, as: https://www.britishwater.co.uk).

[7]   Desalination: A National Perspective. National Academies Press, 2008. (Accessed 20 December 2023, as: nap.nationalacademies.org/catalog/12184/desalination-a-national-perspective).

[8]   International Desalination Association. Web site. (Accessed 20 December, 2023, as https://idadesal.org).

[9]   Singapore Water Association. Web site. (Accessed 20 December, 2023, as https://www.swa.org.sg).

[10] Aquaver. Website. (Accessed 20 December 2023, as: hugedomains.com/domain_profile.cfm?d=aquaver.com).

[11] GE. Website. (Accessed 20 December 2023, as: https://www.ge.com/news/press-releases/ge-expands-aftermarket-services-ultrafiltration-membrane-bioreactor-reverse-osmosis).

[12] Lenntech. Website. (Accessed 20 December 2023, as: https://lenntech.com).

[13] Cryodesalination: Oceans of free water. Website. (Accessed 20 December 2023, as: https://cryodesalination.com).

[14] Water softener solutions. Website. (Accessed 20 December 2023, as: watersoftenersolutions.com/water-and-process-how-water-softeners-work/).

[15] United States Geological Survey, Mineral Commodity Summaries, 2023. Website. (Accessed 18 December 2023 as: https://www.usgs.gov, https://doi.org/10.3133/mcs2023, as a downloadable pdf).

11 Iron and steel

11.1 Introduction

The production of iron has a history that reaches back into ancient times. Some Egyptian tombs contain small iron beads, tools, or weapons, even though no iron ore is mined in or around the Nile River. This ancient iron was refined from meteorites that were scavenged from the area in and around ancient Egypt [1]. The Roman Empire may be the first that produced iron on a large scale, generally for weapons, tools, and construction. A fascinating report in 1994 based on Greenland ice cores showed conclusively that Roman working of iron was done on a large enough scale that it polluted much of western Europe, including the ice of Greenland [2]! This gives an indication both of how large the scale of iron production could be using nothing more than the technology of ancient times, as well as how much pollution iron refining can generate. But no iron production from past times rivaled that which began as part of the Industrial Revolution. This is when the production of this metal began to transform from many small operations to the large furnaces that are used today.

There are a number of different companies producing iron and steel today, on all six inhabited continents. Currently, ArcelorMittal has become one of the largest steel manufacturing company in the world [3]. It states at its web site:

"Steel is as relevant as ever to the future success of our world. As one of the only materials to be completely reusable and recyclable, it will play a critical role in building the circular economy of the future. Steel will continue to evolve, becoming smarter, and increasingly sustainable." [3]

While such statements are always biased toward the company in a positive fashion, it is obvious that steel possesses qualities that make it an excellent, very durable building material.

11.2 Ore sources

Iron ores are found widely throughout the world. Table 11.1 shows a nonexhaustive list of them, but we must note that most iron is refined from only three of those listed: hematite, magnetite, and taconite. It can be seen that two of these ores have some of the highest iron percentages of those listed.

Prior to the Second World War, taconite ores were not mined, since hematite and magnetite ores were plentiful. But as the sources that were easiest to reach have been mined out, taconite, which is still plentiful, has become an economically feasible source of iron.

https://doi.org/10.1515/9783111329512-011

Table 11.1: Various Iron Ores.

Name	Formula	Percent Iron	Geographic Location	Other Metals in Ore
Ankerite	$Ca(Mg, Mn, Fe)(CO_3)_2$	Varies	Peru	Magnesium, manganese
Goethite	$FeO(OH)$	62.8		
Greenalite	$Fe_4Si_2O_5(OH)_4$	52.3	USA, Minnesota	Both Fe^{2+} and Fe^{3+}
Grunerite	$Fe_7Si_8O_{22}(OH)_2$	39.1	South Africa	
Hematite	Fe_2O_3	69.9		
Laterite	Mixed $Fe_xAl_yO_z$	Varies	India, Australia,	Aluminum, nickel
Limonite	$FeO(OH) \cdot n\,H_2O$	52.3		
Magnetite	Fe_3O_4	72.4		
Minnesotaite	$(Fe, Mg)_3Si_4O_{10}(OH)_2$	30.7	USA, Minnesota	
Siderite	$FeCO_3$	48.2		
Taconite	Fe_3O_4 mixed with quartz	*Usually >15	USA, Minnesota, Michigan,	

*In taconite, the iron is usually present as dispersed magnetite.

11.3 Current iron production

Production of iron by nation is tracked by the USGS in three forms: iron and steel, iron ore, and iron oxide pigments [4]. Figure 11.1 shows the production of iron ore by country. It does not delineate which ores are mined in which countries, but perhaps obviously, all such operations are economically profitable.

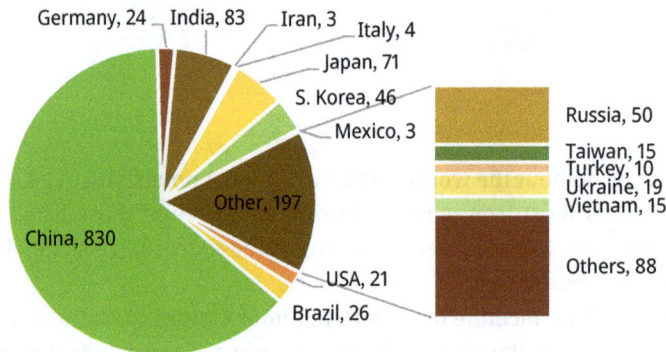

Figure 11.1: Iron ore production (in millions of metric tons).

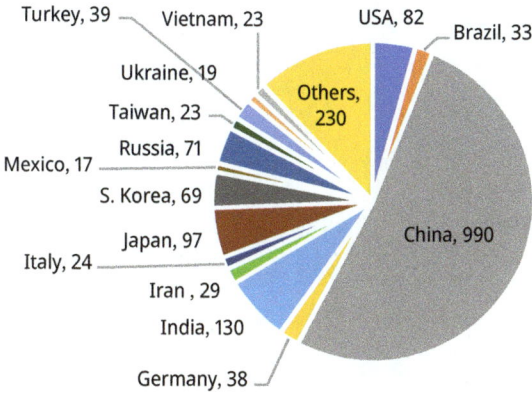

Figure 11.2: Iron and steel production (in millions of metric tons).

The production of iron and steel from various ores is tracked by the USGS Mineral Commodity Summaries [4], the World Steel Association [5], and several other professional trade organizations [6–19]. What is called the apparent steel consumption (or ASP) is also tracked, with China projected to be the largest producer and consumer for the foreseeable future, and India also showing an expected increase.

The production of iron from natural ores is a series of chemical reductions of some iron compound, usually an oxide, to the reduced iron metal, and at the same time the removal of oxygen with some other reducing agent, usually carbon monoxide. The reactions can be represented as a series of three reductions, as follows in Scheme 11.1:

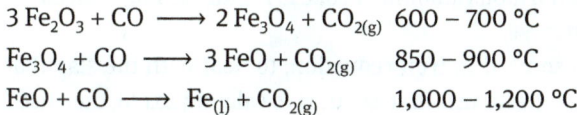

$$3\,Fe_2O_3 + CO \longrightarrow 2\,Fe_3O_4 + CO_{2(g)} \quad 600 - 700\,°C$$
$$Fe_3O_4 + CO \longrightarrow 3\,FeO + CO_{2(g)} \quad\quad 850 - 900\,°C$$
$$FeO + CO \longrightarrow Fe_{(l)} + CO_{2(g)} \quad\quad 1{,}000 - 1{,}200\,°C$$

Scheme 11.1: Reaction chemistry for iron production.

In each step, CO acts as the reducing agent, oxidizing to CO_2. The carbon monoxide is added to the system as coke, meaning a reduced form of carbon, and combusted via an air blast. The simplified reaction for this is

$$2\,C + O_{2(g)} \longrightarrow 2\,CO_{(g)} \quad 200 - 700\,°C$$

Ultimately then, a great deal of coke is required in the refining process, as it is used in stoichiometric amounts in each phase of the reduction. Even accounting for the fact that other metals as well as silicates are present in the original ores, and that these may

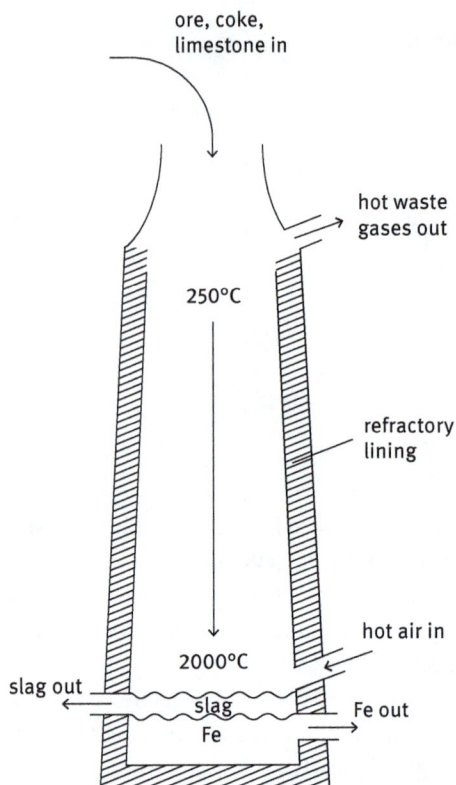

Figure 11.3: Blast Furnace.

not require carbon monoxide, an enormous amount of coke is required, simply because modern refining operations are so large.

A great deal of limestone is also used in iron reduction, to deal with the slag that invariably forms. In two reactions, limestone is first calcined to calcium oxide, then used as a flux to capture SiO_2. The reactions are

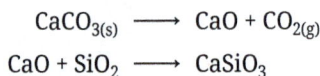

$$CaCO_{3(s)} \longrightarrow CaO + CO_{2(g)}$$
$$CaO + SiO_2 \longrightarrow CaSiO_3$$

As mentioned, what continues to be staggering about the production of iron is the sheer size of these operations. Blast furnaces can now produce over 80,000 short tons of iron weekly, with 48 companies producing raw steel [4]. The diagram of a basic blast furnace that produces iron is shown in Figure 11.3.

Perhaps obviously, the decision to construct and begin operation of a blast furnace, or the decision to decommission one, is one that involves a large number of corporate personnel at the highest levels. Once a blast furnace begins operations, those operations continue ceaselessly, often for years, until the decision is made to stop production.

Table 11.2: Common Types of Steel.

Steel Type	Alloying Element(s)	Uses or Physical Properties
Carbon steel	C	Hardness
High speed steel	W	High hardness
High strength low alloy steel	1.5 % Mn	Greater strength
Low alloy	Mn, Cr, <10 %	Increased hardness
Manganese steel	Mn, 12 %	Wear resistance and durability
Stainless steel	Cr 11 %, Ni	Corrosion resistance
Tool steel	W, Co	Increased hardness, drills & cutting tools

11.4 Steel production

Almost all refined iron, roughly 98 %, is used in the production of some form of steel. Iron is the only element used on an industrial scale in which a less pure form has more desirable physical properties than the rigorously pure form – since steel is often 1–4 % carbon. There are some uses for iron that remain, such as pots, pans, and other end user items. But most iron is alloyed with some other metal or nonmetal to produce one of a wide variety of steels. The World Steel Association indicates that there are more than 3,500 different types of steel, with some claiming only small, niche markets [5]. But clearly, any material that exists in this many different formulas must have an extremely wide variety of uses. Table 11.2 is a brief listing of some of the most common types of steel.

Almost all steel is produced via two techniques: the basic oxygen furnace (BOF) and the electric arc furnace (EAF). Figure 11.4 shows a diagram of a basic oxygen furnace.

It should be noted that a BOF operation is hot enough that it is run in a refractory-lined steel container. As well, the oxygen lance is critical for introducing oxygen gas into the reaction vessel.

An electric arc furnace can also be used to produce batches of steel. A simplified design is shown in Figure 11.5. Note that at the base of any such unit there must be some means of tipping the furnace to pour out the molten metal.

The electric arc furnace method tends to use iron scrap as its principal feedstock. The basic oxygen furnace is more often used with iron ore, coke, and limestone [6].

11.5 Uses of iron and steel

Any exhaustive list of the uses of iron and steel would be enormous. The USGS breaks use down into broad categories, stating, "Construction accounted for an estimated 46 % of domestic shipments by market classification, followed by transportation (predominantly automotive), 26 %; machinery and equipment, 8 %; energy, 6 %; appliances, 5 %;

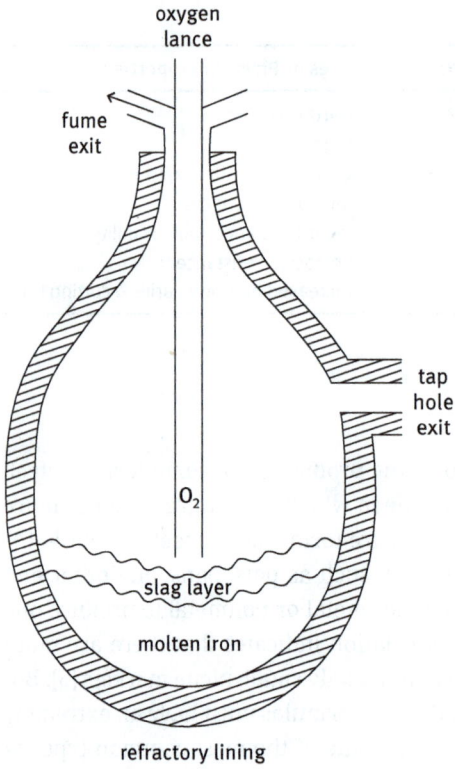

Figure 11.4: Basic oxygen furnace.

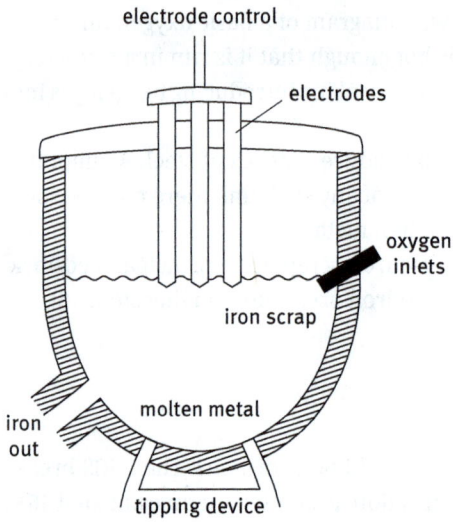

Figure 11.5: Electric arc furnace.

Figure 11.6: Iron Transport via Highway.

and other applications, 9 %." [4] Most people are familiar with the uses of steel in producing cars, bridges, and the reinforcing beams and rods in buildings. Additionally, the many national and regional trade organizations that are devoted to the promotion and regulation of steel list its uses on their web sites [5–19].

To get finished iron and steel to major users, it can be transported by highway, rail, or ship. Unless there is a special need, it is not usually transported by air because of its high density. Figure 11.6 illustrates an example of this, showing a truck on a US interstate highway transporting rolls of iron sheet. Notice that even though there is room for more iron on the truck, the density of it is such that it would ruin the axles on the trailer if the trailer was filled, and possibly damage the roads over which it traveled. Even with two rolls loaded, note that the trailer uses eight axles to support the weight.

The use of iron and steel is so widespread that it is worth noting the commentary given in the USGS Mineral Commodity Summaries about substitutes for these two materials:

> "Iron is the least expensive and most widely used metal. In most applications, iron and steel compete either with less expensive nonmetallic materials or with more expensive materials that have a performance advantage. Iron and steel compete with lighter materials, such as aluminum and plastics, in the motor vehicle industry; aluminum, concrete and wood in construction; and aluminum, glass, paper, and plastics in containers." [4]

11.6 By-product production

Slag has traditionally been considered as one of the major by-products in the production of iron. Enough of it is produced with iron that there is actually an association devoted to finding uses for it, so that it may be considered a co-product in lieu of a by-product. Today, much of it is used in concrete mixes [20].

Carbon dioxide is definitely a by-product of iron and steel production, but for many years was not considered as such, simply because it is an odorless, colorless gas that does not exist as an unsightly material. Yet the emission of enormous amounts of carbon dioxide to the atmosphere is a problem of large-scale waste generation.

To illustrate the amount of carbon dioxide by-product produced during iron production, the overall reaction for iron refining can be used, as shown in Scheme 11.2.

$$Fe_2O_{3(s)} + 3\,CO_{(g)} \longrightarrow 2\,Fe_{(l)} + 3\,CO_{2(g)}$$

Scheme 11.2: Overall Reaction for Iron Reduction from Ore.

A stoichiometric example starting with 1 ton of iron oxide can show how much carbon dioxide is released as a by-product, as follows:

$$1 \text{ metric ton } Fe_2O_3 = 2{,}000 \text{ pounds} \times 453.59\,g/1\,lb = 907{,}180\ g\ Fe_2O_3$$
$$907{,}180\ g\ Fe_2O_3 \times 1 \text{ mol } Fe_2O_3/159.687\,g = 5{,}681.0 \text{ mol } Fe_2O_3$$
$$5{,}681.0 \text{ mol } Fe_2O_3 \times 3 \text{ mol } CO_2/1 \text{ mol } Fe_2O_3 = 17{,}043 \text{ mol } CO_2$$
$$17{,}043 \text{ mol } CO_2 \times 44.010\,g\ CO_2/1 \text{ mol } CO_2 = 750{,}061\,g\ CO_2$$

Converting back to tons:

$$750{,}061\,g\ CO_2 \times 1\,lb/453.59\,g = 1{,}653.6\,lb\ CO_2$$
$$1{,}653.6\,lb \times 1 \text{ ton}/2{,}000\,lb = 0.827 \text{ ton } CO_2$$

Thus, for every 1 ton of iron produced, 0.827 tons of CO_2 is co-produced. Note that the two pie charts in Figures 11.1 and 11.2 are measured in millions of metric tons, which means that this number, 0.827, can be multiplied by the totals in Figures 11.1 and 11.2 to give an estimate of the amount of CO_2 produced as a by-product of iron production.

Rather obviously, these figures do not include all the slag and other impurities that are co-produced in an iron manufacturing process. For example, since most iron ores are not pure iron oxides, there is some sulfur present in the ore as well. This routinely is oxidized to a mixture of SO_2 and SO_3, often called "SO_x", which must either be captured by gas scrubbers, or which can be released to the surrounding environment.

11.7 Recycling

Undoubtedly, iron and steel are recycled in almost every country on a very large basis, and compete with the recycling of paper, glass, and plastic in the view of the general public. This is done widely enough that there are trade organizations devoted to such recycling [20–22]. The cost of producing virgin iron and steel from ore is many times higher than is melting and recycling the metal.

Scarp yards deal with most iron and steel recycling that starts with smaller batches of material. Very few scrap yards deal exclusively with these two metals, as most also are in the business of recycling aluminum, copper, and several other metals. Larger scrap yards may deal exclusively in older automobiles and trucks, which means they also must concern themselves with the nonmetal components of such vehicles. Almost all the components of an automobile can be recycled, although there is approximately 50 kg of what is called automotive shredder residue (ASR) left per vehicle.

The recycling of large iron and steel vehicles and structures, such as decommissioned ocean-going cargo ships or steel bridges, is usually done with some direct contract between the firm doing the de-construction and another firm which has the blast furnace capability to handle the incoming material.

Currently, efforts have been undertaken to determine how CO_2 generated in iron production can be captured and used in some other way. This is technically not recycling, but represents a means of waste reduction as long as the end result is less CO_2 being released to the atmosphere. These efforts have not yet resulted in a process that can be scaled up to an industrial-sized operation.

Bibliography

[1] 5,000 years old Egyptian iron beads made from hammered meteoritic iron. T. Rehren, et al. *Journal of Archaeological Science*, Volume 40, Issue 12, December 2013, Pages 4785–4792.
[2] Hong, Sungmin; Candelone, Jean-Pierre; Patterson, Clair C.; Boutron, Claude F. (1994): "Greenland Ice Evidence of Hemispheric Lead Pollution Two Millennia Ago by Greek and Roman Civilizations", Science, Vol. 265, No. 5180, pp. 1841–1843.
[3] ArcelorMittal. Website. (Accessed 21 December, 2023, as: https://corporate.arcelormittal.com). XXX Baosteel Group. Website. (Accessed 21 December 2023, as: https://www.baosteel.com/.
[4] United States Geological Survey, Mineral Commodity Summaries, 2023. Website. (Accessed 18 December 2023 as: https://www.usgs.gov, https://doi.org/10.3133/mcs2023, as a downloadable pdf).
[5] World Steel Association. Website. (Accessed 21 December 2023, as: https://worldsteel.org).
[6] World Steel Organization. Website. (Accessed 21 December 2023, as: https://worldsteel.org/steel-topics/statistics/annual-production-steeldata/).
[7] American Iron and Steel Institute. Website. (Accessed 21 December 2023, as: https://www.steel.org).
[8] Institute of Materials, Minerals, and Mining. Website. (Accessed 21 December 2023, as: https://www.iom3.org/iron-steel-group.html).
[9] The Iron and Steel Institute of Japan, ISIJ. Website. (Accessed 21 December 2023, as: https://www.isij.or.jp/english/index.html).
[10] Japan Iron and Steel Federation. Website. (Accessed 21 December 2023 as: https://www.jisf.or.jp/en/).
[11] Eurofer, the European Steel Association. Website. (Accessed 21 December 2023, as: https://www.euofer.eu).
[12] European Confederation of Iron and Steel Industries. Website. (Accessed 21 December 2023, as: Bloomberg.com/profile/company/0611171D:BB).

[13] Korea Iron & Steel Association. Website. (Accessed 21 December 2023, as: https://www.kosa.or.kr/sub/eng/about/sub01.jsp).

[14] African Iron and Steel Association. Union of International Associations. Website. (Accessed 21 December 2023, as: https://uia.org//s/or/en/1100056539).

[15] South African Iron and Steel Institute. Website. (Accessed 21 December 2023, as: https://saisi.co.za).

[16] Australian Steel Institute. Website. (Accessed 21 December 2023 as: https://www.steel.org.au).

[17] British Stainless Steel Association. Website. (Accessed 21 December 2023 as: https://bssa.org.uk).

[18] International Nickel Study Group. Website. (Accessed 21 December 2023 as: https://insg.org).

[19] International Stainless Steel Forum. Website. (Accessed 21 December 2023 as: https://www.worldstainless.org).

[20] National Slag Association. Website. (Accessed 21 December 2023, as: https://nationalslag.org).

[21] Steel Recycling Institute. Website. (Accessed 21 December 2023 as: https://www.nerc.org).

[22] Steel Recycling Locator. Website. (Accessed 21 December 2023, as: https://steel.org/sustainability).

12 Aluminum

12.1 Introduction and history

Aluminum is one of only a few materials that have moved from the point of discovery to major industrial use in only 60 years. As a metal used on an industrial scale, aluminum is large enough that there are now several companies that produce aluminum and aluminum-based chemicals as their major products [1–4], and also several organizations that advocate its uses [5–7].

First isolated in 1827, aluminum is so difficult to extract from its ore that from that time until 1859, when the Deville process for the production of alumina (aluminum oxide) was developed, it qualified as a rare metal. The Deville process (sometimes also named the Deville–Pechiney process) does not produce aluminum metal, but does produce pure alumina, from which the metal can be more easily extracted. This was entirely superseded in the early twentieth century by the Bayer process. The last Deville process works were in Germany in the 1930s.

12.2 Bauxite sources

While there are several aluminum-bearing ores, bauxite is the one that is the most economically feasible to mine and from which to extract aluminum; and thus all aluminum is refined from it. Table 12.1 lists the major aluminum-bearing ores. Bauxite usually appears as a red rock, the red color being the result of iron oxides in the mineral. Depending on the ore source, bauxite can be as low as 30–35 % aluminum oxide, with the balance being iron oxides, titanium dioxide, or other silicates.

Bauxite is not evenly or particularly widely distributed throughout the Earth's crust. The United States Geological Survey tracks bauxite mining for aluminum production, and notes in its annual Mineral Commodities Summary that the US is 100 % dependent

Table 12.1: Aluminum-bearing ores.

Name	General Formula	Locations	Comments
Bauxite	Mixture of Al minerals	All six inhabited continents	Primary source of Al ore, generally 30–50 % Al_2O_3
Boehmite	γ-AlO(OH)	France	
Diaspore	α-AlO(OH)	USA, Russia, Hungary, Turkey	
Goethite	α-FeO(OH), α-AlO(OH)	France, Germany	Iron-bearing ore
Gibbsite	$Al(OH)_3$	USA	
Hematite	Fe_2O_3, Al_2O_3	USA	Iron-bearing ore

https://doi.org/10.1515/9783111329512-012

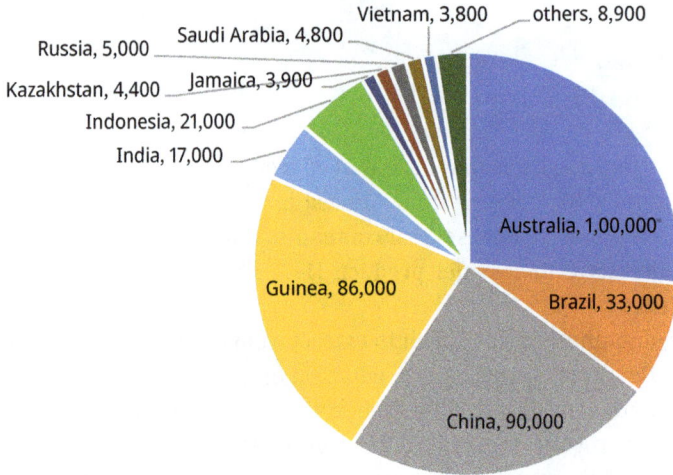

Figure 12.1: Bauxite production (in thousands of metric tons) [8].

upon imports for its bauxite [8]. Countries that mine the mineral include Jamaica, Brazil, Guinea, and Australia. The global production of bauxite is shown in Figure 12.1.

12.3 Aluminum production methods

Since alumina is refined using the Bayer process, and is the feedstock for the Hall–Heroult process, we will discuss this first, and then the production of aluminum metal.

12.3.1 Alumina production

The Bayer process is the method by which alumina is produced on an industrial scale. Largely, it is a chemical means of separating iron oxides and other impurities from the aluminum oxide by means of solubilities at different temperatures. Scheme 12.1 shows the basic reaction chemistry by which this is done.

The first reaction not only produces the sodium aluminum oxide, which is soluble, but takes advantage of the difference in its solubility with that of the much less soluble iron oxides, titanium dioxide, and silicates that are present in the ore (although not shown in the first reaction). The resulting insoluble residues are called "red mud", and are separated. The end product in Scheme 12.1 appears to be exactly the same as the

$$Al_2O_{3(impure)} \longrightarrow 2\,NaOH \longrightarrow 2\,NaAlO_2 + H_2O \qquad \approx 180\,°C$$
followed by
$$NaAlO_2 + 2\,H_2O \longrightarrow Al(OH)_3 + NaOH$$
followed by
$$2\,Al(OH)_3 \longrightarrow Al_2O_3 + 3\,H_2O_{(g)} \qquad\qquad at \approx 1{,}000\,°C$$

Scheme 12.1: Bayer process for alumina production.

starting material, but that is because the focus is on the production of the aluminum-containing portion of the ore. The product is now free of impurities.

12.3.2 Aluminum refining

The process by which aluminum metal is refined from ore is called the Hall–Heroult process, patented by both Charles Hall in the United States and Paul Heroult in France. This process brought the price of aluminum metal down roughly to where it is today, moving it from a very expensive commodity to one that is considered quite cheap. In simplified terms, the reduction of aluminum from alumina can be shown as in Scheme 12.2.

$$2\,Al_2O_3 + 3\,C \longrightarrow 4\,Al_{(l)} + 3CO_{2(g)}$$

Scheme 12.2: Hall–Heroult process aluminum refining.

The reaction definitely involves more than the addition of carbon, which functions as the anode, but can be simplified to the above. It must be noted that the alumina is actually in solution with a molten cryolite (Na_3AlF_6) bath, which means that the reaction chemistry involves aluminum fluorides and oxo-fluorides as the reduction occurs. Figure 12.2 shows the schematic of a cell in the Hall–Heroult process. Note that the position of the anode is such that when it is worn it can be replaced with relative ease. Also, CO_2 is formed as the reaction proceeds, and must be captured or vented. As well, HF is generated, and must be captured.

Molten aluminum sinks to the bottom of each Hall–Heroult cell, because in its molten or liquid state it is denser than molten cryolite. This is the reverse of their densities when they are solids. Because of this, liquid aluminum must be vacuum siphoned from the cells, which involves breaking through the upper crust that forms as the reaction proceeds. The aluminum that is removed is cast into ingots, and fresh alumina is added to each cell.

Aluminum refining also requires a large input of electricity. For this reason, smelters are often co-located with an electricity generating plant, although power can

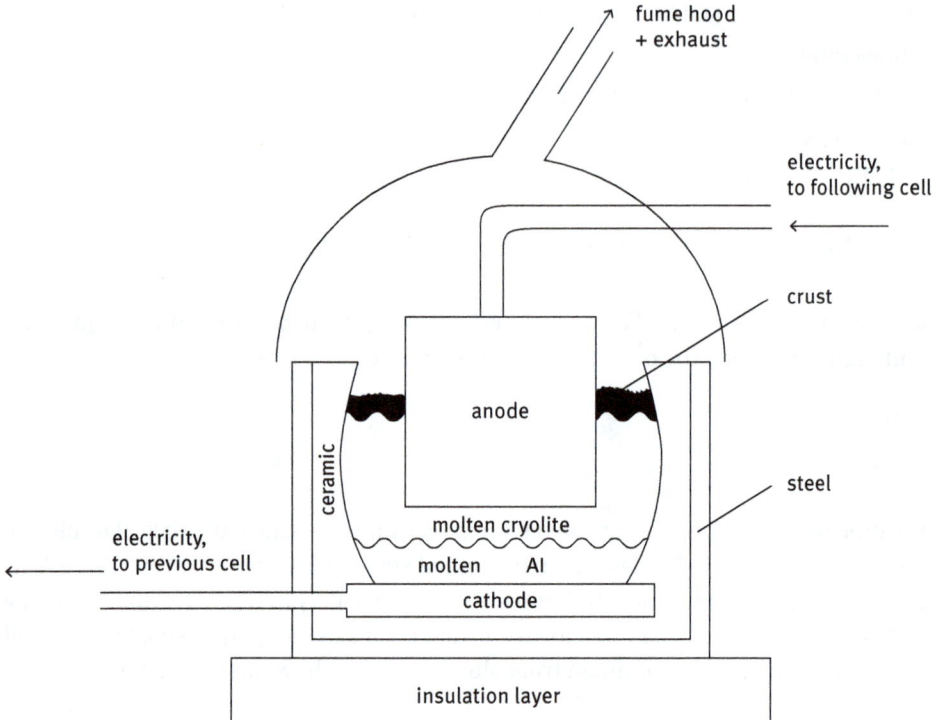

Figure 12.2: Hall–Heroult process cell.

be transferred to the smelter location. The recently opened smelter in Hafnarfjordur, Iceland, which is operated by Rio Tinto Alcan was so located because that country has an inexpensive abundance of geothermal and hydroelectric power. It has proven economically profitable to ship the ore and cryolite to the site, and refine the metal there. The company web site notes the economic impact of the plant when it states, "The company plays a big role in Iceland's economy and supplies about 23 % of all the goods exported from Iceland it produces 202,000 tonnes of some of the highest quality aluminium billets with the lowest carbon footprint in the world. 100 % of our electricity is generated from clean, renewable hydropower, supplied by the National power company Landsvirkjun" [3].

12.4 Major industrial uses

Aluminum plays a major role in modern society in applications in which low density and light weight are important. Consumer-end products are numerous, and while beverage cans and airplane skins are probably thought of most readily by the general public as

Table 12.2: Use of aluminum.

Use	Example
Building and construction	Building panels, window frames, doors, mobile home parts
Consumer goods	Refrigerators, kitchen appliances, cookware, air conditioning equipment
Containers	Cans, foils, chemical containers
Electricity	Power lines, towers, structural supports
Machinery	Heat exchangers, chemical equipment
Transportation, parts for:	Aircraft, small boats, cars, light trucks, trailers, shipping containers
Other	Recreational boats, camp gear, eating utensils

uses of aluminum, there are many others as well. Table 12.2 is a non-exhaustive list of applications that currently utilize aluminum.

12.5 Lightweight alloys

Aluminum has become a metal with numerous major industrial uses because of its low density. Whereas iron has a density of 7.87 g/cc, aluminum has a density of only 2.70 g/cc. There are hundreds of aluminum alloy formulations, all based on some industrial need. Aluminum–silicon alloys and aluminum–scandium alloys have found multiple uses in the aerospace industry, and some alloys have even been named, as brands can be associated with such names. For example, "Titanal" is an aluminum alloy of Austria Metal AG, which is used extensively in making skis and snowboards [9]. But this is only one of many aluminum alloys, and only one of several that have trade names.

The American National Standards Institute (ANSI) lists numeric codes to the many aluminum alloys that have found some industrial use or niche use, as well as to many other metals and their alloys [10]. The International Alloy Designation System has developed series, starting with the 1000 series, all of which designate different alloying elements with aluminum. The 6000 series designator is generally for silicon and magnesium. The 7000 series is alloyed with zinc, and normally includes alloys that have very high strength to weight ratios. The 8000 series can cover numerous elements, but include aluminum–lithium alloys, which tend to be quite light [10]. Some common alloys and their numbers are listed in Table 12.3.

While a large set of numbers in a table can be confusing at first glance, close observation of Table 12.3 shows some trends. First, the total amounts of zinc are decidedly higher in the 7000 series than the 6000 series. As well, the percentages of magnesium tend to be higher in the 7000 series. Additionally, the amount of copper is higher in the 7000 series, with one exception.

Table 12.3: Examples of Aluminum alloys (in percentage per element).

Alloy #	Cu	Cr	Fe	Mg	Mn	Si	Ti	Zn	Al
6061	0.40	0.35	0.70	1.20	0.15	0.80	0.15	0.25	Remainder
6066	1.20	0.40	0.50	1.40	1.10	1.80	0.20	0.25	Remainder
6070	0.40	0.10	0.50	1.20	1.00	1.70	0.15	0.25	Remainder
6162	0.20	0.10	0.50	1.10	0.10	0.00	0.10	0.10	Remainder
7005	0.10	0.20	0.40	1.80	0.70	0.35	0.06	5.00	Remainder
7068	2.40	0.05	0.15	3.00	0.10	0.12	0.01	8.30	Remainder
7075	2.00	0.28	0.50	2.90	0.30	0.40	0.20	6.10	Remainder
7178	2.40	0.28	0.50	3.10	0.30	0.40	0.20	7.30	Remainder

12.6 By-products

The red mud listed in the Bayer process contains a significant amount of iron, which has posed large environmental and disposal challenges in the past. The material is highly alkaline, and must be dried to minimize its environmental impact. Recently, Vedanta Aluminum, Ltd, has undertaken a large-scale operation at its plant in Odisha, India to refine red mud, reclaim much of the sodium hydroxide, and market the resulting red mud powder for use in cement mixes and other industrial applications [4].

12.7 Recycling

In the last chapter, we mentioned how much and how often iron and steel are recycled. While aluminum does not rival iron is the sheer mass of it that is recycled (because of aluminum's very low density), in terms of percentage used and recycled, aluminum and iron are recycled to roughly the same levels. Once again, it requires much more energy to refine aluminum from a bauxite ore than it does to reclaim the refined metal. Many states and municipalities world-wide have programs whereby aluminum cans are recycled. Cans are the most obvious consumer end use product. But aluminum from many other applications is recycled as well, often through scrap yards.

Bibliography

[1]   Alcoa. Website. (Accessed 21 December 2023, as: https://www.alcoa.com).
[2]   Rio Tinto. Website. (Accessed 21 December 2023, as: https://www.riotinto.com).
[3]   Rio Tinto Alcan Iceland. Website. (Accessed 21 December 2023, as: https://www.riotinto.com/en/operations/iceland).
[4]   Vedanta Aluminum, Ltd. Website. (Accessed 21 December 2023, as: https://vedantaaluminum.com).
[5]   The Aluminum Association. Website. (Accessed 21 December 2023, as: https://www.aluminum.org).
[6]   European Aluminum. Website. (Accessed 21 December 2023, as: https://european-aluminum.eu).
[7]   Australian Aluminium Council. Website. (Accessed 21 December 2023, as: https://aluminium.org.au).

[8] United States Geological Survey, Mineral Commodity Summaries, 2023. Website. (Accessed 18 December 2023 as: https://www.usgs.gov, https://doi.org/10.3133/mcs2023, as a downloadable pdf).

[9] Austria Metall, AG. AMAG. Website. (Accessed 21 December 2023, as: https://www.amag-al4u.com/en/products/neuerscheinungen).

[10] American National Standards Institute. Website. (Accessed 21 December 2023, as: https://www.ansi.org).

13 Copper

13.1 Introduction

Copper is one of the elemental metals that has been known and worked since ancient times. Other such metals include: gold, silver, iron, lead, and tin. The island nation of Cyprus is so named because of the copper mined from it over two millennia ago. Indeed, the alloying of copper and tin to make bronze (copper and zinc alloyed together makes brass) has been known for so long that many civilizations which have discovered how to make this material are said to have gone through a "Bronze Age" [1]. For example, such an age for the Indus Valley civilization, now in Pakistan, occurred over 4500 years ago. This is evidenced by bronze artworks such as "the Indus Dancing Girl", a cast bronze figure discovered on a dig at Mohenjo-daro in 1927. As well, the functional and artistic bronze objects created during the Han Dynasty in ancient China show such skill that they have become valuable items in the world's art markets today (so much so that experts must be on constant watch for modern forgeries). Likewise, *Archaeology* magazine states in its January/February 2015 issue:

"Beginning about 1700 B.C., a new material became available in northern Europe that would change the way entire classes of objects were made and how wealth and status were expressed. In Denmark, bronze, which was imported through extensive trade networks with southern Europe, became the material of choice for tools, weapons, ceremonial objects, and jewelry for more than a thousand years" [2].

Routinely, a bronze age occurs before an iron age in a civilization because the melting temperature of iron is higher than that of bronze, and thus it requires some greater source of heat to obtain the temperatures needed to work iron. Throughout history, coal-fed fires have been used to produce bronze.

Also throughout history, iron tools and weapons have tended to displace those made of bronze because of their ability to hold a sharp edge longer, and thus do the job or task for which they were made for a longer time. A specific example is the manufacture of swords. Bronze swords longer than approximately $2'$ (approximately 60 cm) can be bent during heavy use, such as battle. Iron swords do not bend or deform at this or even greater length, and thus can be made more than a foot longer (over 91 cm total). But although people favor iron tools and thus displaced those based on copper, copper mining, refining, and use underwent a major expansion as the Renaissance and Industrial Revolution came to pass in Europe. The need for copper in equipment as mass armies were fielded, followed by the need for copper as wiring was used first in telegraph lines, and later in an ever-expanding electrical and telephone grid, meant that more and more copper was required.

Clearly, the use of copper and copper-bearing alloys made huge differences in the lives of peoples from ancient times to the present, and played a part in advancing several civilizations [1–3].

https://doi.org/10.1515/9783111329512-013

13.2 Ore sources

Copper is mined widely throughout the world, and can be mined as both reduced metal and as various ores. Some copper mineral deposits form extremely attractive crystals, and thus have been of interest to jewelers as well as amateur mineral collectors. Table 13.1 lists various copper ores and their approximate copper percentage by weight, as well as geographic locations where they have been extracted.

Unlike some metals, such as aluminum, several different copper-bearing ores can be refined profitably. The USGS Mineral Commodity Summaries tracks the worldwide production of copper, but does not do so for any one specific ore. Rather, it tracks "Ores and concentrates" that are imported and exported, as well as "Copper from all old scrap." The world mine production of copper is shown in Figure 13.1, measured in thousands of metric tons [4].

Table 13.1: Copper ores and sources.

Ore name	General formula	Approximate % Cu (based on the formula)	Geographic location/comments
Azurite	$2\,CuCO_3 \cdot Cu(OH)_2$	55.2	France
Bornite	$2\,Cu_2S \cdot CuS \cdot FeS$	63.3	Widely distributed
Chalcocite	Cu_2S	79.9	Highly profitable Cu source
Chalcopyrite	$CuFeS_2$	34.5	Canada, South America
Chrysocolla	$CuO \cdot SiO_2 \cdot H_2O$	38.0	Congo, Indonesia, USA
Covellite	CuS	66.5	Russia, USA
Cuprite	Cu_2O	88.8	Bolivia, Chile, France, Russia, USA
Malachite	$CuCO_3 \cdot Cu(OH)_2$	57.3	Congo, Mexico, Russia, USA, Zambia
Tetrahedrite	$(Cu, Fe)_{12}Sb_4S_{13}$	32–44	Germany

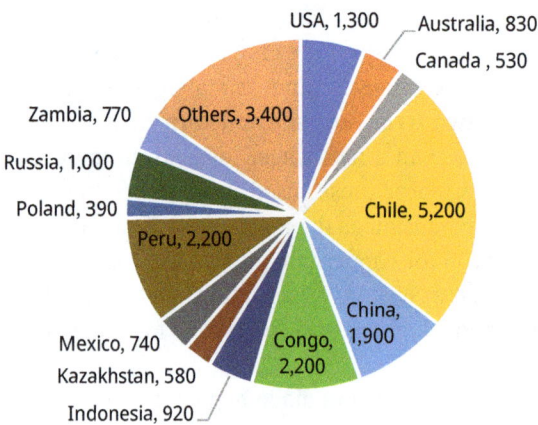

Figure 13.1: World copper production.

Table 13.2: Copper refining companies (copper in thousands of metric tons).

Company	Copper	Other products
Codelco	1,523	Silver, gold, energy
Aurubis AG	1,130	Lead, gold, silver, platinum group metals
Glencore Xstrata Plc	1,100	Nickel, zinc, ferro alloys
Freeport-McMoRan Copper & Gold Inc.	1,080	Gold, molybdenum, cobalt, oil and natural gas
Jiangxi Copper Co.	1,021	Silver, gold, selenium, tellurium, rhenium, sulfuric acid
Grupo Mexico SAB	684	Molybdenum, silver, gold, lead, zinc
Tongling Nonferrous Metals Group Co.	645	Gold, silver, brass, copper sulfate, nickel sulfate
JX Nippon Mining & Metals Corp.	620	Semiconductors, indium
BHP Billiton Ltd.	574	Aluminum, manganese, nickel, iron ore, potash
KGHM Polska Miedź SA	573	Precious metals, rhenium, lead

There are numerous companies which refine copper, with the top 10 being listed in Table 13.2 [5]. None of these companies refines copper exclusively, as can be seen from the table. Copper refining is often associated with that of silver and gold, as well as other rare, expensive metals. This is because such metals are often recovered from what is called "anode mud" that is produced in the copper refining process.

There are certainly other industrial concerns that produce copper, but these ten produce a significant percentage of the world's output annually [6–15]. Because there are so many firms mining and refining copper, several organizations exist that promote the use of copper as well as of its alloys [16–23].

The price of copper is also tracked on world commodity markets, and has hovered between approximately US$ 3.30 per pound and US$ 4.25 per pound for the past few years.

13.3 Production methods

Discussing the production of copper metal in terms of reaction chemistry is difficult because there are so many ore sources, as listed in Table 13.1, some of them rich in oxygen and others rich in sulfur. These different ores require different treatments to effect the reduction of copper from some cationic state to its reduced form. But some refining steps can be said to be common to most ores. They include the following:
1. Milling, crushing, and grinding. This first step is a physical means of getting the ore batch to uniform particle size for further chemical and physical treatment.
2. Concentration. Copper ores must be separated and concentrated from other materials that co-exist with it naturally, such as silicate minerals. After milling, even when

particle size is such that the material is a powder, the starting material may have only 1 % copper in it. The concentration step can bring this up to 10 % and remove insoluble waste materials.

3. Froth floatation. This can be used to separate materials based on their density in water. It is not used in all copper refining operations, but was patented in 1921 precisely for use in refining copper. The First World War had increased the need for the metal on the part of governments on both sides of the conflict, and this in turn drove innovations in ore refining.

4. Leaching. Weak acid solutions can be used to produce copper sulfate solutions, and to separate out metal oxides that are insoluble.

5. Electrowinning or electrolytic processing. Pure copper cathodes and slightly less pure copper anodes are used to accumulate pure copper at the cathode and separate out other metals.

Additionally, if necessary:

1. Smelting. This is the process of melting the reduced metal, often multiple times, to increase its purity. Levels of up to 99 % can be achieved in this manner. Sulfide impurities can be removed at this stage by introducing oxygen and driving off the sulfur as SO_2 gas.

2. Electrolytic refinement. This technique is designed to deposit copper at the cathodes of a refining chamber, much like electrowinning, and can build resultant cathodes that weigh up to approximately 125 kg.

3. Cathode conversion. This final step involves shaping the refined metal, ultimately for sale to users and intermediate customers who use the metal as a feedstock for further production of alloys and end user items.

Perhaps obviously, there will be differences in these processes based on different ores, or even based on different batches of material.

Copper of extremely high purity is refined by the above electrolytic method, in which a highly pure copper cathode is placed in solution with a copper anode of slightly lower purity. Passing current between the two allows migration of copper atoms from the anode to the cathode, and ultimately the loss of the anode. When controlled properly, this process leaves behind insoluble metals which often include those that have higher values per weight than copper, such as silver, gold, and platinum. Known as "anode mud", this material is almost always recovered and separated to extract the metals. Depending on the starting batch of copper, the recovery of metals such as gold from anode muds helps defray or eliminate the cost of the electricity needed in the system. Figure 13.2 shows a schematic representation of this process.

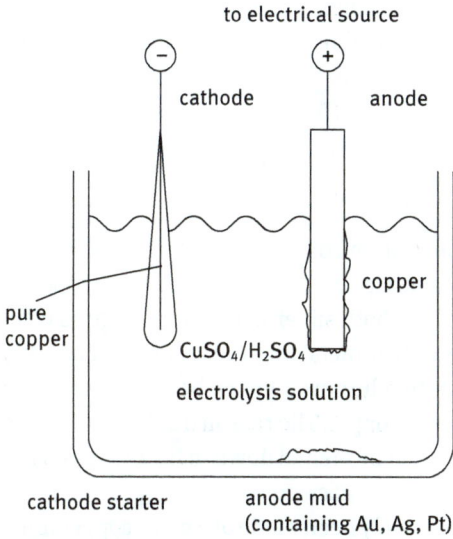

Figure 13.2: Electrolytic refining of high purity copper.

13.4 Major industrial uses

Undoubtedly, many people think of copper wiring as the major use of copper metal in industrial, commercial, and residential settings. But this is only one among many. Some of the large scale uses include the following:

13.4.1 Wiring

Aluminum competes with copper when it comes to wiring, but in many cases, copper wiring is preferred due to its known performance capabilities. Copper is quite ductile, and can be drawn into wiring that spans a large array of thicknesses. The American Wire Gauge (AWG) system has been in use for over 100 years to determine the diameter of wires, both copper and others. This is important in being able to determine how much current a particular gauge of wire can carry safely.

13.4.2 Piping

Although copper piping competes with various forms of plastic piping, it still consti-tutes a significant market share for the various applications in which pipes are used. As with copper wiring, copper tubing is a developed enough application that a range of piping of various diameters and thicknesses are marketed, and the Copper Development Organization even makes available a downloadable publication, *The Copper Tube Handbook* [3].

13.4.3 Coinage

In general, people consider the smallest denomination coins, such as cents, Euro-cents, two-Euro-cents, and even half-cents in some countries, to be the major use of copper in coinage. But copper has been used traditionally in both silver and gold coinage as well. For example, the term "coin silver" means 90 % silver metal and 10 % copper metal. This alloy was used in several countries for almost two hundred years, because it is lower than sterling silver, which is 92.5 % silver and 7.5 % copper. The reason this is important is that sterling silverware and other tableware cannot be melted down to make counterfeit coins unless the counterfeiter wishes to take a loss on each coin.

Additionally, circulating gold coins have routinely been made of a gold–copper alloy because the resulting coins are harder than they would be if made of pure gold, and the more recent gold bullion coins made by several national mints may contain copper as well, again so that the gold is hardened and does not wear easily. For example, the South African Krugerrand is a gold coin that contains 1.00 troy ounces of gold, but with a total mass of 1.09 troy ounces also contains 0.09 troy ounces of copper. In terms of percentage this is 91.67 % gold and 8.33 % copper. In more traditional terms, this is 22 carat gold. As well, the United States Gold Eagle coins, which have been minted since 1986 as one-ounce coins, are 91.67 % gold, 3.00 % silver, and 5.33 % copper.

In terms of circulating coins, many nations use copper in some alloy in numerous denominations that do not appear to be copper. The United States nickel, or five-cent piece, is one example, with an alloy of 75 % copper and only 25 % nickel metal. The circulating coinage of Great Britain has many more examples, shown in Table 13.3.

Table 13.3: Compositions of current coins of Great Britain.

Denomination	Copper (%)	Other Metal (%)	Comments
2 pound	76	Zn – 20, Ni – 4	A two ring coin with two different colored alloys
1 pound	70	Zn – 24.5, Ni – 5.5	
50 pence	75	Ni – 25	
20 pence	84	Ni – 16	
10 pence	75	Ni – 25	In 2011, switched to 94 % steel, 6 % Ni
5 pence	75	Ni – 25	In 2011, switched to 94 % steel, 6 % Ni
2 pence	97	Zn – 2.5, Sn – 0.5	In 1992, changed to 94 % steel, 6 % Cu
1 pence	97	Zn – 2.5, Sn – 0.5	In 1992, changed to 94 % steel, 6 % Cu

It can be seen from Table 13.3 that even for such coins as the 5, 10, 20, and 50 – pence pieces, which have a silver look to them, that a large amount of copper is utilized.

13.5 Brass

The term "brass" generally means an alloy of copper and zinc, although there are many different alloys that qualify as brass, almost all of which can include further elements. Like bronze, brass has been known and used for centuries, and useful alloys were often found through trial and error. Different brasses find use in many functional end products, as well as in many decorative ones. Table 13.4 shows a partial listing of brass alloys that have been found to be useful in one application or another.

The uses of some of the brass compositions are perhaps obvious, such as that for cartridge brass for ammunition in weapons and sporting firearms. The name "tombac" has become common among jewelers and others who work in the brass industries, as the name for a class of brasses that have an attractive look [3]. Others are less common, such as the Mn–brass, which was apparently made by the United States Mint in response to

Table 13.4: Brass compositions, in percent.

Name	Cu	Zn	Al	Fe	Mn	Ni	Pb	Sn	Use
Admiralty brass	69	30						1	Various
Aich's alloy	60.66	36.58		1.74				1.02	In seawater environments
Alpha brass	65	<35							
Beta brass	50–55	45–50							Can be cast
Cartridge brass	70	30							Ammunition casings
DZR brass	>95								Small amount of As
Gilding metal	95	5							Ammunition
High brass	65	35							Rivets and screws
Low brass	80	20							Metal adapters
Mn-brass	70	29			1.3				US dollar coins
Muntz metal, or duplex brass	60	40		<1					In seawater environments
Ni-brass	70	24.5				5.5			1 pound coins in UK
Nordic gold	89	5	5					1	10, 20, and 50 Euro cent coins
Prince's metal, or Prince Rupert's metal	75	25							
Red brass	85	5					5	5	
Rivet brass	63	37							
Tombac	85	15							Jewelry
Yellow brass	67	33							

complaints by the vending machine industry that all vending machines throughout the nation would have to be re-fitted if the golden dollar coins were made with a different electric signature than that of the previously issued Susan B. Anthony dollars (all coins dropped into a vending machine have an electric current passed quickly through them as a deterrent to using counterfeit coins or slugs).

13.6 Bronze

Bronze is another alloy of copper, this time of copper and tin, although there are many compositions for modern bronzes that have small amounts of other elements as well. Table 13.5 gives a selective listing of several bronze alloys that have found some industrial use.

As mentioned, the ability to make and use different bronzes goes back to ancient times. Indeed, for many civilizations the first "age" that comes after a "Stone Age" is often called "The Bronze Age." Bronze has proven to be a very useful material because it can be made on a small scale ultimately to produce farm tool and weapons.

To produce different bronzes, much like brasses, copper is melted, usually at temperatures of 550 °C–650 °C, and tin is added, as well as other elements that give a particular bronze its characteristics. The molten material is then poured into molds to make ingots, from which other objects can be fashioned.

The C" numbers in Table 13.5 are part of the Unified Numbering System (UNS), which lists the standards of the different bronze alloys, and correlates to what particular attributes each alloy possesses. Note how closely one alloy can be to another, yet be categorized by a different UNS C" number.

Table 13.5: Various Bronzes.

Type/Family	Cu	Sn	P	Zn	Fe	Mn	Al	Pb	Comments
Phosphor bronze	97.5–98.5	1.0–1.7	0.03–0.35	0.30	0.10			0.05	C50500
Phosphor bronze	Remainder	4.2–5.8	0.03–0.35	0.30	0.10			0.05	C51000
Phosphor bronze	Remainder	7.0–9.0	0.03–0.35	0.02	0.10			0.05	C52100
Manganese bronze	58.5	1.00		39.2	1.00	0.30			
Aluminum bronze	92.0						8.00		

13.7 Other copper alloys

Besides brass and bronze, there is still a wide variety of alloys which require copper. Some may use copper as a minor component, such as pewter. Others may use small amounts of elements not normally considered to be part of alloys, such as beryllium, but still be called brasses or bronzes. There are enough copper alloys that the UNS has been developed to encompass all of them. Several examples of different copper alloys that are not generally considered brass or bronze are listed in Table 13.6 [24].

Table 13.6: UNS designators for copper alloy examples.

UNS No.	Major Elements	Comments
C15000	Cu, Zr	Used for welding
C15760	Cu, Al	Good cold-working properties
C16200	Cu, Cd	High conductivity and tensile strength, Cu-99 %, Cd-1 %
C17200	Cu, Be	Aerospace and oil drilling applications, Cu-98 %, Be-2 %
C17510	Cu, Ni, Be	High strength alloy
C18000	Cu, Ni, Cr, Si	High hardness, heat treatable
C18150	Cu, Cr, Zr	High electrical conductivity

13.8 By-products

The refining and production of copper, as well as of copper-based alloys, has produced a large amount of waste over the course of years. The production of copper for small change – one cent coins, or pennies – in the United States has left behind a waste pond in Montana that is extremely toxic, and that has been the topic of at least one critical article [24].

Before the advent of modern pollution control systems, the refining of copper, and the use of large amounts of coal to smelt it, led to some heavily polluted areas in terms of slag and sulfur oxide co-production. For example, Swansea, Wales produced more than half of the western world's copper in the mid-1800s, but at an environmental cost that made local cynics claim that, "…if the Devil were to pass that way he would think he was going home" [3, 25, 26]. Perhaps obviously, copper refining has undergone drastic improvements, to ensure the environmental footprint of such industry is minimized.

13.9 Recycling and re-use

Copper is very often recycled and reclaimed in scrap yards throughout the world. The USGS Mineral Commodity Summaries for 2013 states, "Old (post-consumer) scrap, converted to refined metal, alloys, and other forms, provided an estimated 160,000 tons of copper in 2022, and an estimated 670,000 tons of copper was recovered from new (manufacturing) scrap" [4]. Copper is recycled to the same extent as other industrially useful metals such as iron and aluminum.

Bibliography

[1] *Sixty Centuries of Copper* by B Webster Smith, UK Copper Development Association in 1965.
[2] "Artifact", *Archaeology*, January/February 2015, p. 68.
[3] Copper Development Association, Inc. Website. (Accessed 22 December 2023, as: https://www.copper.org).
[4] United States Geological Survey, Mineral Commodity Summaries, 2023. Website. (Accessed 18 December 2023 as: https://www.usgs.gov, https://doi.org/10.3133/mcs2023, as a downloadable pdf).
[5] Trade Street Digest. Website. (Accessed 22 December 2023, as: https://wwwtradestreetdigest.com/2023-copper-mining-report).
[6] Codelco. Website, (Accessed 22 December 2023, as: https://www.codelco.com).
[7] Aurubis AG. Website, (Accessed 22 December 2023, as: https://www.arubis.com).
[8] Glencore Xstrata Plc. Website, (Accessed 22 December 2023, as: https://www.glencore.com).
[9] Freeport-McMoRan Copper & Gold Inc. Website, (Accessed 22 December 2023, as: https://www.fcx.com).
[10] Jiangxi Copper Co. Website, (Accessed 22 December 2023, as: https://www.en.jxcc.com).
[11] Grupo Mexico SAB. Website, (Accessed 22 December 2023, as: https://www.gmexico.com).
[12] Tongling Nonferrous Metals Group Co. Website, (Accessed 22 December 2023, as: https://www.bloomberg.com/ Tongling Nonferrous Metals Group Shanghai International).
[13] JX Nippon Mining & Metals Corp. Website, (Accessed 22 December 2023, as: https://www.jx-nmm.com/english).
[14] BHP Billiton Ltd. Website, (Accessed 22 December 2023, as: https://www.bhp.com).
[15] KGHM Polska Miedź SA. Website, (Accessed 22 December 2023, as: https://kghm.com).
[16] Copper Development Association, Inc. Website. (Accessed 22 December 2023, as: https://www.copper.org).
[17] American Copper Council. Website, (accessed 22 December 2023, as: https://www.americancopper.org).
[18] Copper Alliance. Website, (accessed 22 December 2023, as: https://copperalliance.org).
[19] International Copper Study Group. Website, (accessed 22 December 2023, as: https://icsg.org).
[20] Copper and Brass Supply Chain Association. Website. (Accessed 22 December 2023, as: https://www.copper-brass.org).
[21] Bronze.Net. Website (accessed 22 December 2023, as: www.bronze.net).
[22] International Lead and Zinc Study Group. Website, (Accessed 22 December 2023, as https://www.ilzsg.org).
[23] National Mining Association. Copper's Versatile and Essential Role. Website. (Accessed 22 December 2023, as: https://nma.org/2021/12/06/coppers-versatile-and-essential-role-2?).
[24] Cadi Company, Inc. Website. (Accessed 22 December 2023, as: https://www.cadicompany.com).
[25] Pennies from Hell: In Montana, the bill for America's copper comes due. Web site (accessed 22 December 2023, as: http://harpers.org/archive/1996/10/pennies-from-hell/).
[26] Alexander, W.O. Development of the Copper, Zinc and Brass Industries in Great Britain from A.D. 1500 to 1900 *Murex Rev.* (1955), 1, (15), p. 399.

14 Other major metals for industrial use

We have discussed iron, steel, aluminum, and copper in Chapters 11–13, but there are several other metals the extraction and refining of which enable many of the processes and capabilities of our modern, technologically advanced, interconnected civilization. Some have been known and used since antiquity, while others have much more modern histories and specific uses. Here we discuss several of them, how they are extracted, and how they are used.

14.1 Titanium

Titanium is an element well known for its toughness and light weight – the latter meaning low density. Although it was discovered in 1791, it was little more than a laboratory curiosity and phenomenon before William J. Kroll left his native Luxembourg in 1938, and patented what is now called the Kroll process for titanium refining in 1940 [1]. All industrial scale titanium production has occurred since that time, and the metal is now produced on a large enough scale and is used in enough vital industries that the United States Geological Survey tracks it in its annual Mineral Commodity Summaries [2]. This extractive process remains an energy intensive one, and thus recycling of titanium can be profitable.

14.1.1 Sources

Titanium ores are mined in several countries, and because of the difficulty of obtaining the metal from its ores, in numerous cases it is not refined to the reduced metal, but rather to titanium dioxide. Titanium dioxide is itself a useful commodity with several end user applications. In the United States, titanium is mined in both Utah and Nevada. Worldwide, China has become a major producer of titanium in the past two decades. Figure 14.1 shows titanium mining, with the United States omitted, because mining firms there consider their production numbers to be proprietary [2].

The reaction chemistry that illustrates titanium refining, called the Kroll process, can be expressed as follows, in Scheme 14.1.

This reaction chemistry is quite simple, and disguises the complexity of the operation. It is routinely performed at 1,100 °C. As well, titanium tetrachloride must be manufactured, often from ilmenite ores, because titanium dioxide cannot be directly reduced. Scheme 14.2 shows the simplified reaction chemistry by which ilmenite (shown as $FeTiO_3$) is converted to titanium tetrachloride. This too is a high-temperature reaction, generally run at 900 °C.

https://doi.org/10.1515/9783111329512-014

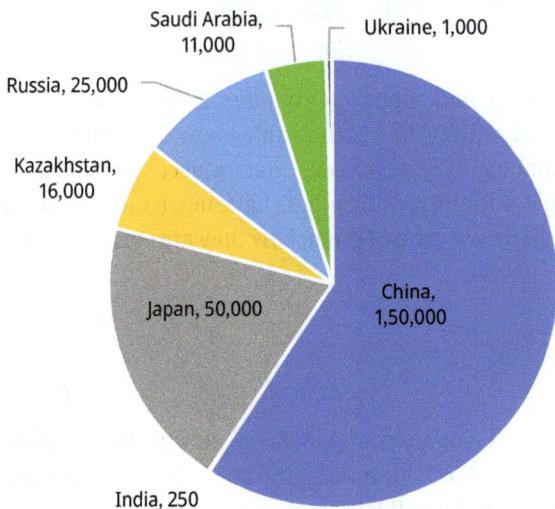

Figure 14.1: Titanium production, in metric tons.

$$TiCl_4 + 2\,Mg \longrightarrow Ti + 2\,MgCl_2$$

Scheme 14.1: Titanium refining reactions.

$$2\,FeTiO_3 + 6\,C + 7\,Cl_{2(g)} \longrightarrow 2\,TiCl_4 + 6\,CO_{(g)} + 2\,FeCl_3$$

Scheme 14.2: Titanium tetrachloride production.

As can be seen in the reaction, titanium production becomes a large scale use of elemental chlorine, as well as of elemental carbon. Chlorine production and the chlor-alkali process were discussed in Chapter 7.

14.1.2 Titanium dioxide production

As mentioned, titanium is energy intensive to extract from its ores. Additionally, titanium dioxide is a useful titanium-bearing material itself, and thus is isolated for use without reduction to the metal. The reaction chemistry for the two processes by which titanium dioxide is refined and isolated, the ilmenite process and the Chloride process, are shown in Schemes 14.3 and 14.4.

The ilmenite process has proven to be economically favorable when the initial concentration of titanium is low. Sulfuric acid is required for the separation of iron and titanium. While this becomes a major use for sulfuric acid, this application is still much smaller than the use of sulfuric acid for the production of phosphoric acid, discussed in Chapter 2.

$2 H_2SO_4 + FeO \cdot TiO_2 \longrightarrow FeSO_4 + 2 H_2O + TiOSO_4$

$2 H_2O + TiOSO_4 \longrightarrow H_2SO_4 + TiO_2 \cdot H_2O$

$TiO_2 \cdot H_2O \longrightarrow TiO_2 + H_2O$

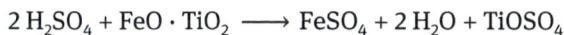

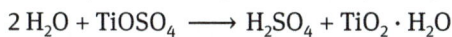

Scheme 14.3: The ilmenite process for titanium dioxide production.

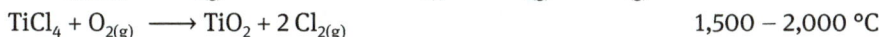

$3 TiO_{2(crude)} + 6 Cl_{2(g)} + 4 C \longrightarrow 3 TiCl_{4(l)} + 2 CO_{2(g)} + 2 CO_{(g)}$ 900 °C

$TiCl_4 + O_{2(g)} \longrightarrow TiO_2 + 2 Cl_{2(g)}$ 1,500 – 2,000 °C

Scheme 14.4: The chloride process for titanium dioxide production.

The Chloride process as shown in Scheme 14.4 does not show the other elements present in a sample of crude titanium dioxide (aka. rutile ore). Trapping the titanium as the chloride permits its separation. The final step of the Chloride process allows the recapture of a large amount of the chlorine.

14.1.3 Uses

The USGS Mineral Commodity Summaries states that nearly three quarters of all titanium produced is used by the aerospace industry, for lightweight alloys. The remainder, "was used in armor, chemical processing, marine, medical, power generation, sporting goods, and other nonaerospace applications." [2] The major uses within the aerospace industry are for lightweight alloys and durable, corrosion-resistant parts. ATI, a major manufacturer of titanium materials, states at its web site: "Titanium and titanium alloys are used in jet engine components, as well as critical airframe applications where high strength and fracture toughness are necessary. Titanium is also used for fasteners and tubing throughout commercial and military aircraft" [3].

Titanium dioxide is used predominantly as a pigment. Once more, according to the USGS Mineral Commodity Summaries, use for titanium dioxide, "paints (including lacquers and varnishes), plastics, and paper. Other uses of TiO_2 pigment included catalysts, ceramics, coated fabrics and textiles, floor coverings, printing ink, and roofing granules" [2]. In the category "other" is the food grade titanium dioxide that is used as a whitener in various candies and chewing gums. When used as a food additive, titanium dioxide is given the number E171.

14.1.4 Recycling

The recycling of titanium is routinely a matter of recycling alloys. Titanium dioxide is not recycled, since it is almost always is consumed in some other process, or is used in some consumer end-use product.

14.2 Chromium

Chromium is another metal with a relatively short history, having first been isolated in 1797. Today the economically useful chromium-bearing mineral is chromite ore, $FeCr_2O_4$. Since chromium is always thus found with iron, the two must be separated.

14.2.1 Chromium sources

Chromium is found in various parts of the world, and is not usually recovered as a by-product of other mining and refining operations. Because of the need for chromium in stainless steel or what are called "superalloys", the United States Defense Logistics Agency tracks production of the element, as does the USGS in their Mineral Commodity Summaries [2, 4]. Figure 14.2 shows where in the world chromium is found and produced. The USGS does not include the United States in these production figures, even though there is an active mine in Oregon, citing proprietary corporate concerns.

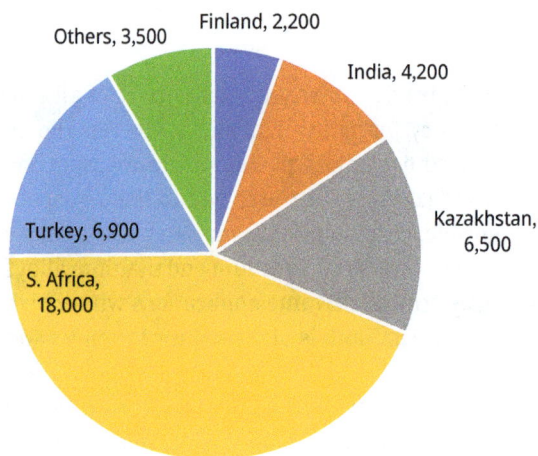

Figure 14.2: Chromium production, in thousands of metric tons.

14.2.2 Uses of chromium

The largest use of chromium is the production of ferrochromium, an iron–chromium alloy that can be higher than 50 % chromium. To obtain the metal, the reduction of chromium from ore must always be accompanied by its separation from iron, which is accomplished with a sodium carbonate and calcium carbonate mixture. The simplified reaction chemistry is shown in Scheme 14.5.

$$4\ FeCr_2O_4 + 7\ O_{2(g)} + 8\ Na_2CO_3 \longrightarrow 8\ Na_2CrO_4 + 8\ CO_{2(g)} + 2\ Fe_2O_{3(s)}$$
$$2\ Na_2CrO_4 + H_2SO_4 \longrightarrow Na_2Cr_2O_7 + H_2O + Na_2SO_4$$
$$Na_2Cr_2O_7 + 2\ C \longrightarrow Cr_2O_3 + Na_2CO_3 + CO_{(g)}$$
$$2\ Al + Cr_2O_3 \longrightarrow 2\ Cr_{(l)} + Al_2O_3$$

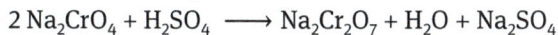

Scheme 14.5: Chromium reduction and isolation.

The first reaction is run at high temperature, and produces the insoluble iron oxide, which makes separation from the chromate possible. Sulfuric acid is then used to produce sodium dichromate, which is then reduced to a chromium(III) oxide through the use of carbon. The final reaction in which chromium is liberated from the oxide by forming an aluminum oxide qualifies as a Goldschmidt reaction, or thermite reaction, in which a molten metal is liberated from a metal(III) oxide by reaction with aluminum powder [5].

Ferrochromium

Ferochromium alloys and stainless steel alloys are important in a number of end-user applications, several of them in military. The USGS Mineral Commodity Summaries states that there is no substitute for the element chromium in stainless steel [2]. Likewise, the US Department of Defense makes similar claims in its annual report on stockpile requirements [6].

Chrome plating

Chrome plating on a variety of metal surfaces has become an increasingly widespread use of chromium for the twin reasons of its resistance to corrosion and its general attractiveness on the end product. Exposed metal automotive components are often chrome plated, although other metals parts can be as well.

Pigment and dye applications

Chromium is used in several pigments because of the bright colors it imparts, usually yellows and greens. Some pigments are derived from mixing two colors to create a third. Examples of chromium-containing pigments are listed in Table 14.1.

Table 14.1: Chromium-containing pigments.

Name	Color	Formula	Comment
Cadmium green	Green	CdS and Cr_2O_3	Mixed pigment
Chrome green	Green	Cr_2O_3	
Viridian	Green, dark	$Cr_2O_3 \cdot H_2O$	
Chrome orange	Orange	$PbCrO_4$ and PbO	Mixed pigment
Chrome yellow	Yellow	$PbCrO_4$	

14.2.3 Recycling

The recycling of chromium is always connected to the recycling of iron and steel materials, especially stainless steel. Almost one-third of chromium consumed annually can come from recycled materials.

14.3 Mercury

Mercury has both an ancient history and an ancient name: quicksilver. It has been known for centuries that the metal can be used to amalgamate gold and thus separate it from other metals and the surrounding soil and silicates in which it is found. It has been used in the production of hats in the past – and long-term inhalation of the vapor resulted in a phrase that is still used today on occasion: mad as a hatter. And while the ill-health caused by mercury inhalation is well known, there are still some parts of the world in which small-scale gold mining involves a miner vaporizing a pan of mercury–gold amalgam in an open fire as a means of separating the mercury from the desired gold, while holding the pan in his or her hands.

14.3.1 Sources of mercury

The mineral cinnabar is the chief ore of mercury, from which the metal can be extracted profitably. There are other mercury-containing minerals as well, such as livingstonite ($HgSb_4S_8$) and corderoite ($Hg_3S_2Cl_2$), but these have not proven profitable for mercury extraction.

The trade in previously refined mercury is also a large, global one. Thus, while amounts of mercury mined, in metric tons, are shown in Figure 14.3, countries like the United States can be listed as mercury exporters, even though no mercury has been mined within the United States in recent years [2].

While China dominates the world market for mercury production at the present time, Figure 14.3 shows that mercury ores are mined widely throughout the world [2]. Mercury extraction from the ore is somewhat more straightforward than that for other metals, like chromium, and can be represented in a simplified manner as shown in Scheme 14.6.

Heating the crushed ore in the presence of oxygen gas from the air, followed by the condensation of the resulting vapor, results in the liquid form of the metal.

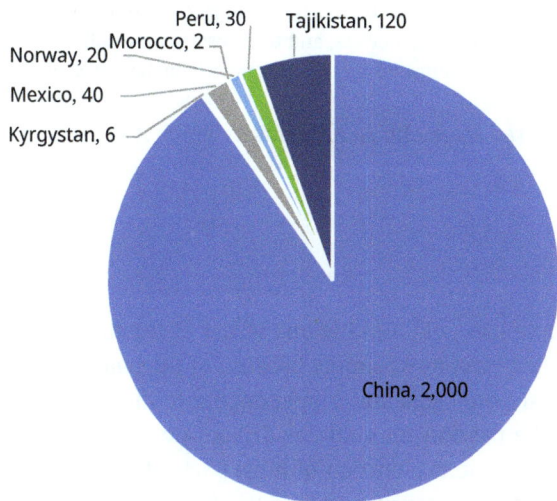

Figure 14.3: Mercury production, in metric tons.

$$HgS_{(s)} + O_{2(g)} \longrightarrow Hg_{(l)} + SO_{2(g)}$$

Scheme 14.6: Mercury isolation.

14.3.2 Uses of mercury

Mercury thermometers are the consumer end use that comes immediately to mind when a person thinks of applications for this metal, but the chlor-alkali process for the production of sodium hydroxide, in which mercury is used as an electrode and during the process amalgamates the sodium, uses far more mercury than any other application. This was discussed in Chapter 7. This has remained the chief use of reduced mercury metal for several decades, although this may decrease in the future, as chlor-alkali plants are no longer being designed to use the mercury system. Rather, plants using membrane separation are being built preferentially.

14.3.3 Recycling of mercury

Mercury is recycled whenever possible. In the United States, there are six companies active in mercury recovery and recycling. Recovered mercury is easily purified, and again sold for any of the uses already mentioned. The USGS Mineral Commodity Summaries makes the following statement about mercury recycling:

"Mercury-containing automobile convenience switches, barometers, compact and traditional fluorescent bulbs, computers, dental amalgam, medical devices, and thermostats were collected by smaller companies and shipped to the refining companies for retorting to reclaim the mercury" [2].

It appears then that mercury is recovered from almost all items in which it is used and then re-used in some way.

14.4 Gold

No other metal seems to excite people like gold. Gold is one of the elemental metals known from antiquity, and valued then (and now) simply because of the beauty of its appearance. Of all the elemental metals, only gold and copper appear to the naked eye as a color other than silver-white. The passion for gold has driven a wide variety of different human explorations, including the exploration of huge parts of the western world. What are now called gold rushes have taken people to the Australian Outback, the American south and west and the Yukon, in the Canadian far north. As well, the search for gold was one of the motivators for the Soviet Gulag prison system to set up camps in eastern Siberia, at or near the Kolyma River.

Although gold is still valued for its appearance, and is used in a wide array of jewelry and other ornamental applications, there are now several industrial uses for the metal and its alloys as well.

14.4.1 Sources of gold

What are called placers or placer nuggets are naturally occurring pieces of gold that are found in the ground or in stream beds as reduced metal. Extremely large placers have been named, and although the general belief is that the day and the age of finding such pieces of gold has passed, occasional finds continue to occur. A find in Victoria, Australia in 2013 was noteworthy because the fields in this area had been previously considered to be no longer economically profitable. Table 14.2 lists some of the largest nuggets found.

While large gold nugget finds are impressive, newsworthy, and able to make people think that the age of buried treasure has not passed, most gold is found in much smaller sizes, often as small as a grain of sand. Such materials require physical and chemical methods of extraction, concentration, and refining.

Figure 14.4 shows where gold is produced today, in metric tons. Although this data has been reported in the USGS Mineral Commodity Summaries, it is not a full accounting for gold consumption, simply because gold is always recycled, and virtually never discarded [2].

Figure 14.4 makes it obvious that gold is widely mined throughout the world. Overall though, it is produced on a significantly smaller scale than metals such as iron, copper,

Table 14.2: Gold nuggets.

Name	Weight (ozt)	Weight (lb)	Location of find	Find Date – Fate
Butte nugget	70	6.07	Butte County, California, USA	July 2014, sold to private collector
Dogtown nugget		54	California, USA	1859
Fricot nugget	201	6.25	California Gold Rush	At California State Mining & Mineral Museum
Hand of Faith	875	27	Kingower, Victoria, AU	1980
Heron	1,008	69.08	Mt. Alexander goldfield	1855
Lady Hotham		98.5	Ballarat, AU	September 8, 1854
Pepita Canaa	1,951	133.80	Serra Pelada Mine, Para, Brazil	September 13, 1983, now at Banco Central Museum
Welcome Nugget	2,218	152.1	Bakery Hill, Ballarat, AU	1858, melted in 1859
Welcome Stranger	2,520	173	Moliagul, Victoria, AU	1869

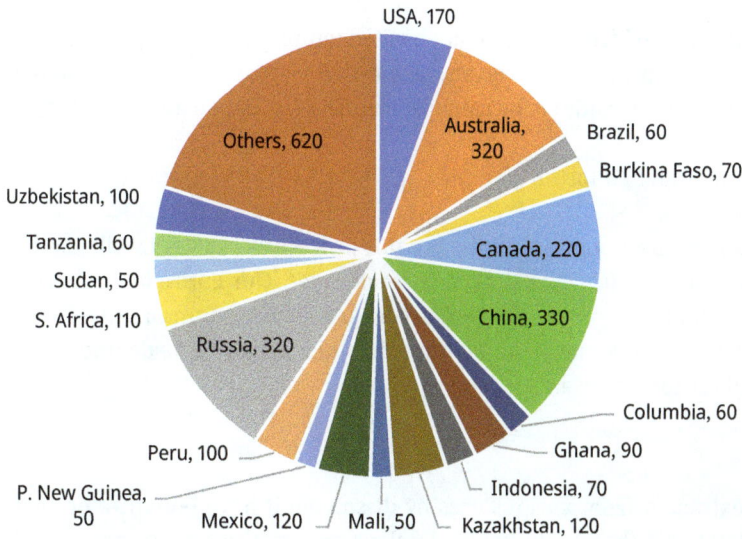

Figure 14.4: Gold production, in metric tons.

and aluminum. Additionally, the numbers shown here do not delineate between gold that is mined and gold that is recovered from anode muds as a by-product from the refining of copper and other metals. This chart also does not take into account the amount of recycled gold that is used each year.

Cyanide leaching

The chemical reaction by which gold is extracted from sands in which the gold exists as tiny particles is shown in Scheme 14.7. This reaction can be run with potassium cyanide or calcium cyanide as well.

$$4\,Au_{(s)(impure)} + 8\,NaCN + 2\,H_2O_{(l)} + O_{2(g)} \longrightarrow 4\,Na[Au(CN)_2]_{(aq)} + 4\,NaOH_{(aq)}$$

Scheme 14.7: Gold extraction with cyanide.

The reaction, called either the cyanide process or less commonly the MacArthur – Forrest process, takes advantage of the affinity of the cyanide ion for gold. Interestingly, this accounts for more than 10 % of cyanide use annually.

The gold-bearing complex must then be reduced back to gold metal. To do this, the Carbon in Pulp (or CIP) method is generally used whenever it is economically feasible. This is difficult to represent with reactions, but can be summarized in a list of steps, as follows:

1. Carbon particles are washed with the gold-bearing complex in solution.
2. The carbon is removed from solution and washed.
3. Gold cyanide is removed from solution at elevated temperature and high pH.
4. The solution then undergoes electrolysis, also called electro-winning. This allows gold to precipitate on the cathode or cathodes, while a less expensive metal is used as the anode.
5. Cathodes are then smelted for further use.

When silver and copper are not mixed in the ore this process works very well, and has been scaled up for industrial use over the last three decades. This entire process is one in which chemically and electrolytically gold is concentrated from an ore in which it may exist as less than 1 % of the material to a state in which it can be made into ingots, and then fashioned for further use.

Aqua regia

Gold can also be extracted from various ores by dissolving it in the mixture of nitric acid and hydrochloric acid that is still known by the common name "aqua regia." This method dissolves the gold and holds it as a complex, $AuCl_4^-$, which is then separated from other materials and reduced back to the metal.

The miller process

This process involves directing a stream of chlorine gas over an impure sample of gold while the metal is being heated. Since gold is less reactive than almost all other met-

Table 14.3: Gold purity.

Carat	Percentage	Comment
24	100	Pure gold, can be scratched with a fingernail
22	91.67	Remainder is usually copper
20	83.33	
18	75	Durable, does not tarnish
14	58.33	Durable alloy, but can tarnish
10	41.67	A hard alloy when the remainder is copper

als, the process forms various metal chlorides, leaving unreacted gold behind. Once the metal chlorides are removed, high-purity gold (as high as 99.9 %) can be isolated.

Anode muds

As mentioned in Chapter 13, gold is recovered from anode muds when copper is refined to high purity. The USGS Mineral Commodity Summaries simply mentions that a "small amount" of gold is recovered in this manner [2].

Gold metal purity continues to be measured in an old system, carats. The system has become well established, with the purity levels of gold jewelry almost always being measured this way. What is called 24 carat gold means essentially pure gold. Several of the other established levels of purity are listed, along with their equivalents in percentage, in Table 14.3.

The second metal in a gold alloy is often copper, although what is called "white gold" can be made by alloying platinum into gold. As well, there are several alloys that are composed of more than two elements. In virtually all cases, when gold is alloyed, it is to harden the resulting metal for whatever use it may undergo.

14.4.2 Uses of gold

Gold continues to be used as a store of wealth and ornamentally, in jewelry. The World Gold Council continuously monitors gold use worldwide, in terms of its uses for storing wealth and in making jewelry, as well as in its rather varied niche industrial uses [7]. Also, the USGS Mineral Commodity Summaries lists the uses of gold, which are shown in Figure 14.5 [2].

The 17 % that is listed as "other" includes ingots and bullion coins. The latter represents a relatively new way for small investors to purchase and own gold. Table 14.4 lists the nations which currently produce gold bullion coins, as well as their weights. It can be seen that several of those listed are sold as coins that are 1/10th or even 1/20th of an ounce, which is affordable even to investors and collectors of modest means.

Gold is also used in small amounts in electronics, especially in electronic connections, because it is highly conductive. For example, the amount of gold used in a cell

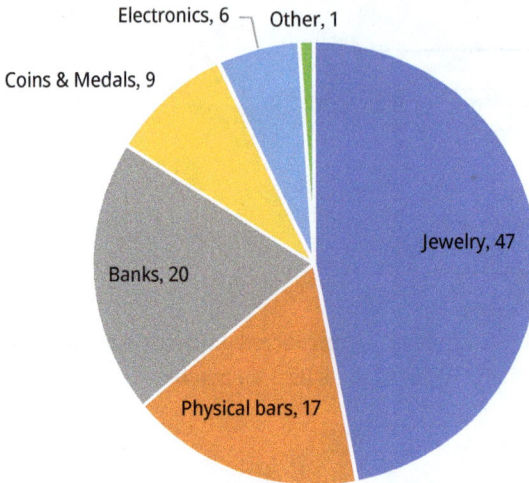

Figure 14.5: Uses of gold.

Table 14.4: Nations issuing gold bullion coins.

Country	Name, gold bullion coin	Gold weight (ozt)	Years issued
Australia	Nugget	1/20, 1/10, ¼, ½, 1, 2, 10, 1 kg	1986–present (larger weights, 1991–present)
Austria	Vienna philharmonic	1/10, ¼, ½, 1	1989–present
Canada	Maple leaf	1/20, 1/10, 1/5, ¼, ½, 1	1979–present
China	Panda	1/20, 1/10 ¼, ½, 1	1982–present
Isle of Man	Angel	1/20, 1/10, ¼, ½, 1	1994–present
Israel	Tower of David	1	2010–present
Kazakhstan	Irbis	1/10, ¼, ½, 1	2009
Malaysia	Kijang emas	¼, ½, 1	2001–present
Mexico	Libertad	1/20, 1/10, ¼, ½, 1	1991–present
South Africa	Krugerrand	1	1967–present
United Kingdom	Britannia	1/20, 1/10, ¼, ½, 1, 5	1987–present
USA	Eagle	1/10, ¼, ½, 1	1986–present

phone has a value of approximately US$ 0.50 (based on a price of US$ 1,200 per troy ounce). But with over a billion cell phones having been produced, this becomes a significant use for the metal.

Additionally, gold as a gold chloride is still used in the production of some red colored types of glass.

14.4.3 Recycling

Gold is routinely recycled simply because of its value. Jewelry stores and dealers in coins and antiquities have routinely purchased gold that people wish to sell, often to raise money. But in the past decade, as the price of gold per troy ounce has risen from US$ 400 per ounce to over US $1,800 per ounce to $2,000, numerous stores have opened which are aimed at purchasing old, unused gold, or gold jewelry that has been in some way broken. The USGS Mineral Commodity Summaries states. "In 2022, an estimated 90 tons of new and old scrap was recycled, equivalent to about 36 % of reported consumption" [2].

14.5 Silver

Like gold, silver has been known and valued since ancient times. Placer nuggets of what are called electrum – a natural mixture of silver and gold – have been found in the rivers of modern-day Turkey since antiquity, and when stamped with an identifying mark served as the world's first coinage. As European expansion brought European explorers and traders into contact with different peoples throughout the world, silver as well as gold was brought back to Europe. This occurred to such an extent in the sixteenth century that by the year 1580, silver from South America had displaced silver from European sources in almost all European commerce. While silver has been mined in many places, historically it was Potosi, Bolivia that produced a huge amount of it for trade in the Americas, China, and Europe, from mines in the mountain often called simply Cerro Rico – Rich Mountain. A multitude of one-, two-, four-, and eight-reales coins, and later Bolivian soles coins made from Cerro Rico silver have been used worldwide, sometimes to the point where they have become almost unrecognizable. Figures 14.6 and 14.7 show,

Figure 14.6: Four soles coin from Potosi, Bolivia.

Figure 14.7: Eight reales coin from Potosi, Bolivia, countermarked multiple times.

respectively, a Bolivian 4 soles and Bolivian–Spanish 8 reales coin, the second of which has been heavily countermarked by Chinese merchants and banking houses as a way of verifying that it was indeed silver metal. The intertwined PTS monogram for Potosi can still be seen in the lower left on each.

14.5.1 Silver production

The profile for the use of silver has changed in the 20th and the 21st centuries, as have the locations and mines from which it is extracted. Coins that are made to circulate are no longer made of silver, although several nations have instituted programs whereby silver coins are made to trade as commodities on the metals markets, much like the gold bullion coins listed in Table 14.4. For this reason, they are usually made in weights of one-ounce or greater.

In the United States, silver is mined primarily in Alaska and Nevada. The USGS Mineral Commodity Summaries records silver production throughout the world in metric tons, as shown in Figure 14.8 [2]. But it should also be noted that silver is mined worldwide as a co-product of the mining of several other metals, including lead, zinc, and copper [8]. The famous mines of Potosi still exist, but now mine tin as the main metal-containing ore.

In Figure 14.8, the category "others" may seem large, but that is because of the just-mentioned mining of silver as a by-product metal. Because of such operations, it appears that silver is widely distributed throughout the world.

The washoe process

One method of concentrating and refining silver, the Washoe process, was discovered and perfected in the 1860s, when silver was being mined from the Comstock Lode, a

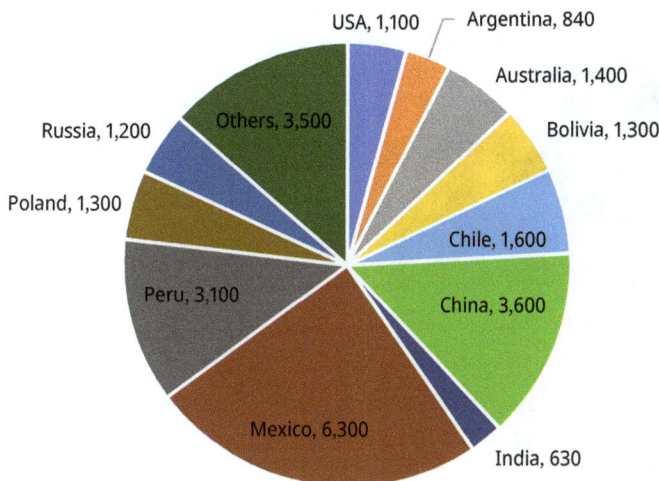

Figure 14.8: World silver production, in metric tons.

famous silver deposit in the western United States that was essentially mined out in the early 20th century. While the process is difficult to show in terms of reaction chemistry, it can be listed in steps as follows:

1. Batches of ore, roughly 550–675 kg each, are crushed to the consistency of sand.
2. The powdered ore is placed in copper pans (iron pans are sometimes used), water is added, and the slurry amalgamated with 2–3 L of mercury.
3. Up to 1.5 kg of sodium chloride and copper sulfate are added to the slurry.
4. Batches are agitated with iron paddles (this provides small but necessary amounts of iron to the mixture).
5. Steam heating and agitation removes impurities.
6. The silver–mercury amalgam is separated from the mixture.
7. The silver is separated from the mercury.

Other methods, such as electro-winning of silver, have become more common in the recent past, and the Washoe process is no longer used in virtually all operations. For lead-silver ores, the Parkes Process is used, in which impure lead with silver in it is added molten to molten zinc. The silver amalgamates in the zinc, and can then be separated.

14.5.2 Silver uses

Because silver is less expensive than gold, it has found a wide variety of uses in industry, and in consumer end products.

The industrial and general use of silver today is shown in Figure 14.9, in terms of percentage of the total use [8].

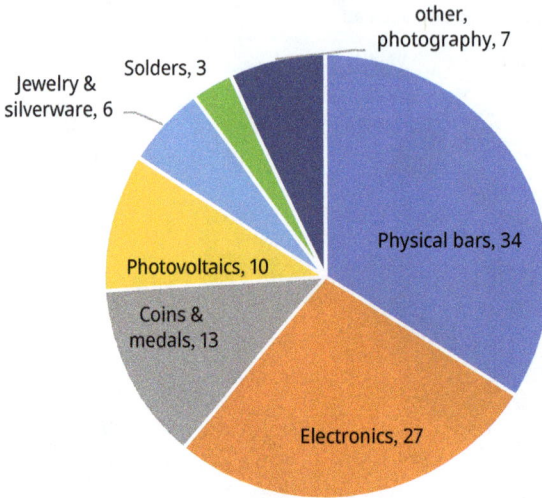

Figure 14.9: Silver use, in percentage.

The rather large "Other" percentage shown here is detailed by the USGS Mineral Commodity Summaries as follows:

"…include use in antimicrobial bandages, clothing, pharmaceuticals, and plastics; batteries; bearings; brazing and soldering; catalytic converters for automobiles; electroplating; inks; mirrors; photography; photovoltaic solar cells; water purification; and wood treatment" [2].

Clearly, silver is used in many ways that involve a small amount of the metal in some user end product, often in some way of which the user is probably unaware. For instance, it is doubtful that most travelers and gamblers know that a tiny amount of silver is used in their passports and possibly in each gambling chip.

Recently, it has been reported that silver is being used along with other materials in 3D printing. While this does not qualify as an industrial scale use of this element at the moment, the rapid and apparently continuing rise in the use of 3D printers for a wide variety of end objects may mean that this will become a significant use of silver in the near future [9].

14.5.3 Recycling

Silver is recycled extensively, with over 1,500 tons recycled in 2012 from various sources [2]. Photography, which has used silver in the formation of the final photograph onto special paper since its inception, uses less silver now than it did 20 years ago because of advances in digital photography and in laser printed images. But old photographs still serve as one source of recycled silver.

14.6 Lead

Lead metal is another metallic element that has been used since ancient time. The lead lining of some waterways constructed during the Roman Empire still exist. Indeed, the word "plumber" and the symbol for the element, Pb, both come from the Latin name for lead, "plumbum."

14.6.1 Lead sources

There is a wide variety of lead-containing minerals and ores, many of which have a high percentage of lead within them. Table 14.5 lists the major lead-containing ores, although it does not describe what other metals may be co-produced from each.

Figure 14.10 shows the sources of lead metal throughout the world, in thousands of metric tons, as reported by the USGS Mineral Commodity Summaries [2]. Nations in Central and South America where lead is mined have also been mined for other, often more valuable, metals as well. For example, both Mexico and Peru have long histories of mining both silver and gold.

14.6.2 Lead extraction chemistry

Lead is often co-produced with another metal, such as silver. When a metal is more valuable than lead, it is the lead which is considered the co-product, yet both metals can be recovered.

The parkes process

This process for separating lead from a silver-containing ore takes advantage of the solubility of zinc with silver, and its immiscibility with lead. Scheme 14.8 illustrates this in a very basic form.

$$Pb:Ag_{(l)} + Zn_{(s)} \longrightarrow Pb_{(l)} + Zn:Ag_{(l)}$$

Scheme 14.8: Lead separation from silver.

The silver can then be isolated by simply heating the resultant alloy until the zinc is driven off as a gas.

Table 14.5: Lead-bearing ores.

Formula	Ore common name(s)	% Lead in ore	Geographic location	Comments
PbO	Lead monoxide, litharge, plumbous oxide, massicot	92.9	USA, England	An uncommon ore; sometimes synthesized when needed.
Pb_3O_4	Lead tetroxide, minium, red lead, triplumbic tetroxide	90.7	Spain	
PbO_2	Lead dioxide, plattnerite, plumbic oxide, scrutinyite	86.7	Europe, Mexico, USA, Russia, Australia, Namibia, Iran	
PbS	Galena, lead(II) sulfide, blue lead, lead glance	86.7	USA, Canada, Germany, Italy, England, Bulgaria, Australia, Israel	Major source of lead for refining, US town named: Galena, Illinois
$PbSO_4$	Angelsite, linarite	68.3	Spain	An uncommon ore
$PbCO_3$	Cerussite, white lead, lead carbonate	77.5	Australia, Germany, USA	Soluble in acids
$PbMnO_4$	Wulfenite, yellow lead	63.5	Austria, USA, Mexico, Slovenia	
$Pb_5(AsO_4)_3Cl$	Mimetite, green lead	69.7	Mexico, Namibia	
$Pb_5(PO_4)_3Cl$	Pyromorphite, green lead	88.9	Australia, Mexico	
$Pb_5(VO_4)_3Cl$	Vanadinite, green lead	73.2	USA, Morocco, Argentina, Namibia	Widely occurring ore

The betterton–kroll process

This is a process by which bismuth and lead are separated. While the reaction is not usually shown stoichiometrically, the reaction chemistry can be represented as shown in Scheme 14.9.

The lead has a higher density and sinks to the bottom of the molten bath in which the reaction is run. Thus, the resulting compounds of bismuth and any other metals are light enough that they can be removed as a dross or slag layer. In general, the reaction chemistry runs at 400–500 °C. If bismuth is being recovered, this can be accomplished by treating this resulting material with chlorine.

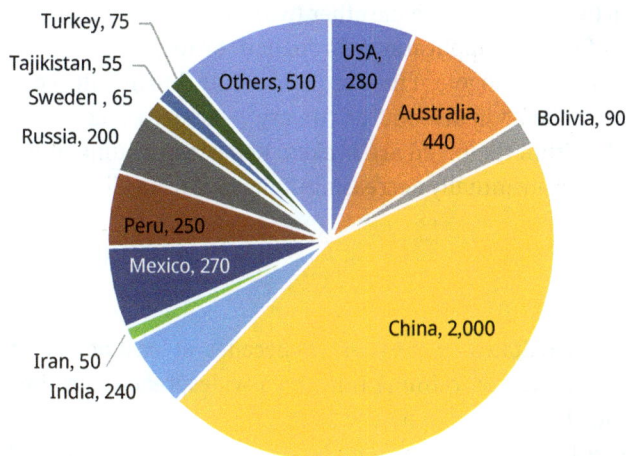

Figure 14.10: Lead production worldwide, in thousands of metric tons.

$$Ca + Mg + Pb:Bi_{(l)} \longrightarrow Ca:Mg:Bi_{(s)} + Pb_{(l)}$$

Scheme 14.9: Betterton–Kroll process for lead isolation.

14.6.3 Lead uses

Lead metal in lead-acid automotive batteries is the single largest application for the metal today, with the USGS claiming that this one use accounts for 92 % of domestic use [2]. But there are numerous other uses, as well. Lead solders are alloys mixed with tin or other metals, although lead-free solders have been claiming a larger market share in the recent past. Lead continues to be used in wheel weights, and among divers, lead continues to find use as a ballast.

14.6.4 Lead recycling

The recycling of lead acid automotive batteries is the major source of recycled lead in the United States. While this is certainly the common source of lead metal, lead from older roofs and building components provides a second source of the metal for recycling.

14.7 Tin

Tin is another one of the elemental metals known from ancient times. Its use in the production of bronze has been discussed in Chapter 13. Tin is one of only a few common metals that can be melted with relative ease, 232 °C, for example, in a wood fire, as opposed to one made from coal (lead and zinc are the other two). This low melting point has made it useful in a variety of end-use applications throughout history.

Additionally, when tin is poured into ingots, if they are sufficiently thin that they can be bent by hand, what has been called "tin scream" or "tin cry" – actually a crackling sound – is heard as the crystals within the metal are broken. Repeated bending of the same tin bar results in the tin scream eventually decreasing or disappearing all together.

14.7.1 Tin sources

Tin has not been mined in the United States for the past two decades, at least not as the primary product. However, tin mining and refining remains a widespread operation. Figure 14.11 shows the worldwide breakdown in metric tons [2].

It can be seen from Scheme 14.8 that tin is essentially mined on all six inhabited continents. While North America is absent from the figure, there are still small amounts of tin that are mined in the greater operations that are located in Alaska.

There are several large corporations that produce tin, as well as an International Tin Research Institute [10]. The top 10 are shown in Table 14.6. These firms are listed alphabetically because production changes annually, and thus the order may change as well [10–21].

Not listed are several companies that produce tin as a secondary product [10].

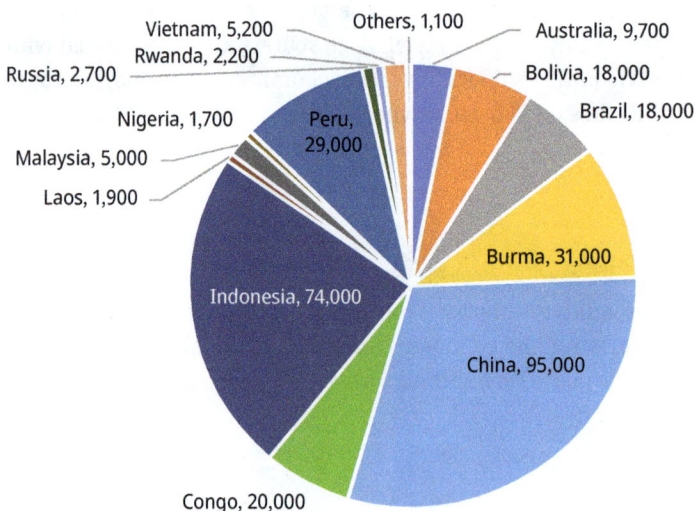

Figure 14.11: Tin production.

Table 14.6: Tin producers, arranged alphabetically.

Company	Location	Tons	Other products
EM Vinto	Bolivia	10,800	
Gejiu Zi Li	China	7000	
Guangxi China Tin	China	14,034	
Malaysia Smelting Corp.	Malaysia	37,792	
Metallo Chimique	Belgium	11,350	Re-processes nonferrous metals
Minsur	Peru	25,399	Lead, copper, antimony
PT Timah	Indonesia	29,600	Lead
Thaisarco	Thailand	22,847	
Yunnan Cheng Feng	China	16,600	Antimony, bismuth, copper, gold, indium, lead
Yunnan Tin Group	China	69,760	

14.7.2 Tin production

Tin refining is a matter of smelting ores at high temperature using coke as a reducing agent. A simplified version of the reaction chemistry is shown in Scheme 14.10.

$$SnO_{2(s)} + C \longrightarrow CO_{2(g)} + Sn_{(l)}$$

Scheme 14.10: Tin refining.

The reaction does not show the limestone or silica that can be used as fluxes, or the fact that the reduction is often run at 1,200 °C to 1,300 °C. Tin(IV) oxide, or tin dioxide, is often mined as the mineral cassiterite. Smelting cassiterite produces tin, the carbon dioxide shown above, and a slag of impurities. As with iron refining, the slag is separated and the molten metal captured. The slag can contain tantalum or niobium, depending on the ore batch, as well as other valuable metals. These can then be recovered and used as well.

14.7.3 Tin uses

The major use of tin that the general public thinks of is cans for the storage of food and other items. This is a significant use of the metal, but there are others as well. The USGS Mineral Commodity Summaries lists uses for tin as follows:

"chemicals, 23 %; tinplate, 22 %; alloys, 11 %; solder, 10 %; babbitt, brass and bronze, and tinning, 7 %; bar tin, 2 %; and other, 25 %."

The use of tin in construction is often for piping and internal components of structures. In terms of transportation, tin finds use in a variety of bearings and other parts that require a softer metal or alloy [2, 21].

Additionally, window glass is currently made by pouring glass onto a pool of molten tin, because the molten tin surface is perfectly smooth, and thus results in glass that has a uniform thickness. This method is often called the float glass process, or the Pilkington process, after its inventor.

Also, tin continues to be used in various alloys that are called pewter [22]. Pewter continues to find numerous niche uses in end items such as flatware, what is called silverware, and other decorative but useful tableware.

14.7.4 Tin recycling

Tin is recycled like other metals, usually through scrap yards. As with most metals, it is much less expensive to recycle tin than to refine ores. Thus, tin recycling is economically driven.

14.8 Platinum group metals (PGM)

The platinum group metals, often abbreviated as PGM, serve a wide variety of purposes in modern industry. The six PGM are: ruthenium, rhodium, palladium, osmium, iridium, and platinum. Some sources now include hassium, meitnerium, and darmstadtium in the PGM group, but the latter three have half-lives that are only seconds long, and have yet to be produced in large enough quantities to find any uses.

14.8.1 PGM sources

The USGS Mineral Commodity Summaries does track the worldwide production of PGM, as it does with many other metals, largely because platinum is produced on a large scale for use in catalytic convertors in the automotive sector. But the PGM are also tracked by the International Platinum Group Metals Association (IPA), which advocates for their use and education concerning them [23]. Figure 14.12 shows the current world breakdown [2].

Platinum was first discovered as a co-product of gold mining in South America during the period of Spanish colonization. Figure 14.11 illustrates how this has shifted with time to South Africa as the main producer of this metal. Platinum is mined from what is called the Merensky Reef, which is part of the Bushveld Igneous Complex.

Palladium production is also spread about the world, and is illustrated in Figure 14.13, again in kilograms [2].

Palladium production is often a by-product metal of other refining. In Russia, it is very often a co-product of nickel refining, while in South Africa, it is a co-product of platinum refining.

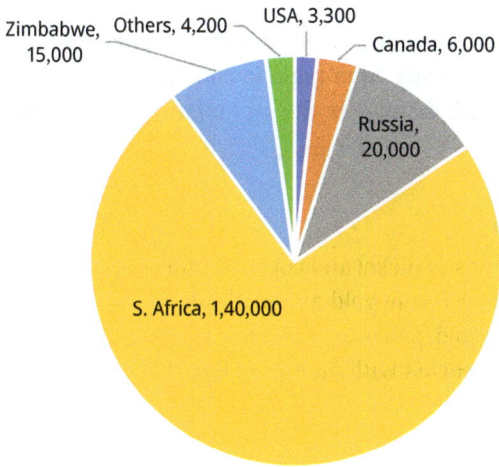

Figure 14.12: Platinum production, in kilograms.

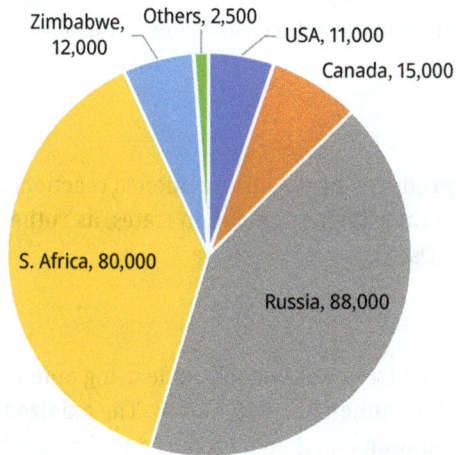

Figure 14.13: Palladium production, in kilograms.

14.8.2 PGM production

Platinum is a remarkably inert metal, and thus its refining has traditionally posed a challenge. Likewise, the entire PGM series is difficult to isolate in some clean, single reaction. Some reactions conditions and comments about the refining of each PGM follows:

Platinum

Platinum can be recovered as a component of anode muds in the refining of copper and nickel, the former of which was discussed in Chapter 13. But it can also be found in placer deposits, usually in the Merensky Reef.

Palladium

Palladium can also be recovered as a by-product of nickel refining or copper refining. Palladium can be separated from platinum using aqua regia (a combination of nitric and hydrochloric acids), since palladium will dissolve in the acid mixture, much more than platinum.

Rhodium

Rhodium is also recovered from anode muds in nickel and copper refining. Ruthenium, osmium, rhodium, and iridium are separated from gold and platinum as well as other metals in the anode mud by dissolving the gold, palladium, and any base metals in aqua regia. These four insoluble PGM are then reacted with molten sodium bisulfate, which separates rhodium.

Iridium

Ruthenium, iridium, and osmium from anode muds that are insoluble in aqua regia, and sodium bisulfate can be reacted with sodium oxide, which separates out insoluble iridium.

Osmium

Osmium and ruthenium salts that have been produced by the just-mentioned reactions, which are water soluble, are then oxidized to their highest oxidation states, as ruthenium tetroxide (RuO_4) and osmium tetroxide (OsO_4).

Ruthenium

Samples of ruthenium tetroxide can be separated from osmium tetroxide using ammonium chloride to produce ammonium hexachlororuthenate, $(NH_4)_3RuCl_6$. The oxidized complexes of metals such as ruthenium and osmium can then be reduced to the metal using elemental hydrogen.

14.8.3 PGM uses

It is difficult to determine a precise listing for the uses of platinum and palladium in terms of amounts or percentages, because the number of automobiles produced each year can change significantly (which means the number of catalytic converters changes), as can the amounts of chemical produced using these two metals as catalysts in catalytic convertors. But a listing of their uses would include:
1. Automotive catalysts
2. Chemical process catalysts

3. Petroleum refining catalysts
4. Laboratory crucibles
5. Computer hard disks
6. Ceramic capacitors
7. Hybridized integrated circuits
8. Jewelry
9. Dental restoratives
10. Bullion coins and exchange-traded funds [2, 24]

The other four elements – rhodium, ruthenium, iridium, and osmium – are used in small enough amounts that even organizations which track the use of PGMs, such as the USGS Mineral Commodity Summaries, do not normally break down the applications for which they are used. There are however numerous small, specialty applications for these four elements. For example, iridium is used to make pen tips for high-end writing instruments, as well as to coat the prongs of engagement rings, because this makes the diamond look clearer. Rhodium is at times used in catalytic converters, and is used as an alloying element for platinum and palladium, because the resulting alloys are harder than either element alone. Ruthenium has found use in solar cells, and also is used to make harder platinum and palladium alloys. Finally, osmium is a component of osmiridium alloys, which are resistant to wear. The compound osmium tetroxide has been used in the past in organic synthesis. It was a useful way to make cis-additions of two oxygen atoms across a double bond. It continues to be used as a staining agent in transmission electron microscopy.

Platinum metal has been marketed in the form of exchange-traded funds (often abbreviated ETFs) and bullion coins, the latter of which provides an investment vehicle in something tangible for the smaller investor. A listing of platinum bullion coins is shown in Table 14.7. Notice that some issues are as small as 1/20th of an ounce of the precious metal. Thus far, only Canada markets a palladium bullion coin.

Table 14.7: Platinum bullion coins.

Country	Coin name	Weight
Australia	Koala	1/20, 1/10, ¼, ½, 1 t.oz.
Australia	Platypus	1 t oz
Canada	Maple Leaf	1/20, 1/10, ¼, ½, 1 t.oz.
Isle of Man	Noble	1/20, 1/10, ¼, ½, 1 t.oz.
USA	Eagle	1/10, ¼, ½, 1 t.oz.

14.8.4 PGM recycling

All of the PGM are valuable enough metals that once they have been refined, they are profitable to recover and recycle. Thus, once again these represent materials the recovery of which is economically driven. As mentioned, platinum and palladium find the most large-scale uses, and so recovery programs are in place for such items as old catalytic convertors. As well, platinum and even palladium jewelry can be re-melted and re-refined if necessary.

14.9 Technetium

Technetium was the last transition metal to be filled in on the modern periodic table. This is one of at least four elements that were hypothesized by Mendeleev, with scandium, gallium, and germanium being the other three. Between the world wars, there was at least one claim of discovery for this element, which was later rescinded. Thus, textbooks from the mid-1930s sometimes include the symbol "Ma" in this spot, for masurium.

All isotopes of technetium are radioactive, although with proper safeguards and protocols, macroscopic amounts of technetium and technetium compounds can be used [25, 26].

14.9.1 Technetium sources and production

Technetium is a nuclear fission waste product, and thus exists in spent nuclear fuel rods, in concentrations that can be greater than 6 %. If needed, technetium can thus be refined and recovered from spent nuclear fuel rods.

A usable form of technetium for medical applications is technetium-99m, or Tc^{99m}, meaning metastable technetium. This decays far too quickly to be recovered from fuel rods, and therefore it is synthesized for short-term applications. To produce this, molybdenum-99 is first produced. It has a half-life of 66 h, and decays to Tc^{99m}, which has a half-life of only 6 h.

14.9.2 Technetium uses

The US Environmental Protection Agency points out that there is no large-scale industrial use for any of the various technetium isotopes, but notes:

"Its short-lived parent, Tc^{99m}, however, is the most widely used radioactive isotope for medical diagnostic studies. Technetium-99m is used for medical and research purposes, including evaluating the medical condition of the heart, kidneys, lungs, liver, spleen, and bone, among others, and also for blood flow studies" [26].

Clearly, this has become an important aspect of modern medicine. But the overall volume of Tc^{99m} that is produced and used remains orders of magnitude smaller than the other metals discussed here.

Elemental technetium that is not the metastable isotope actually does find an industrial use in the production of a few specialty, corrosion-resistant steels. Since all isotopes of technetium are radioactive, these steels are used in what are defined as closed systems, so that there is no danger of radiation poisoning [25].

14.9.3 Technetium recycling and re-use

Technetium is used on a small enough scale that there are no recycling programs for it. Radioactive by-products from technetium use must be disposed of as hazardous waste by licensed waste haulers. The protocols followed for such disposal vary from country to country, and even from one state or municipality to another.

14.10 Tantalum and niobium

Both tantalum and niobium have shorter histories than the metals known from antiquity. The initial discovery of each was somewhat confused, but the generally accepted date of discovery for tantalum is 1802, and that for niobium 1801. Niobium was first called columbium – and indeed still has this name in some metallurgical societies today – because its discovery was from an ore sample that originated in Massachusetts, USA (Columbia being one of the more classical names for North America). Throughout much of the 19th century there was confusion as to whether these were two distinct elements, or if different claims of discovery were for the same element.

These two metals are not found in large enough concentrations that ore sources are mined primarily for them. Like several other metals, these two elements are by-products of other mining operations. These two metals have not been mined in the United States in the past 50 years. Nevertheless, the USGS Mineral Commodity Summaries tracks each of them because they have become vitally important for certain industries [2], and a trade organization exists devoted to them and to increasing general awareness of their uses [27].

14.10.1 Tantalum and niobium sources

Tantalum and niobium are not widely distributed throughout the Earth. Several minerals contain either tantalum or niobium in small amounts, but what is called coltan is the material with the highest concentration of both. Coltan is generally a mixture of two minerals, columbite and tantalite. The first is a niobium mineral, while the second contains a significant amount of tantalum [27, 28].

Tantalum production by country, in metric tons, is shown in Figure 14.14. Note that the production unit is metric tons, and not thousands of metric tons, meaning the overall amount of tantalum produced annually is relatively small when compared to other metals [2].

Niobium is even less widely distributed worldwide than tantalum, with two countries dominating production, as shown in Figure 14.15 [2]. Once again, production is in metric tons.

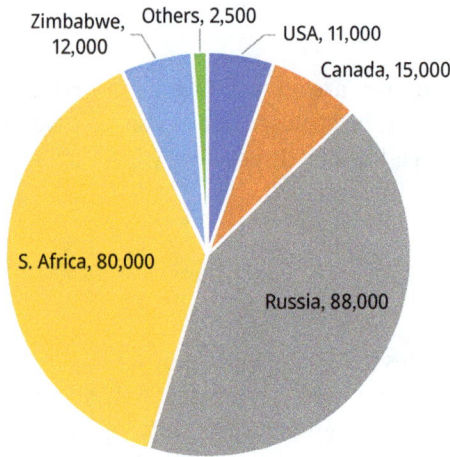

Figure 14.14: Tantalum production worldwide.

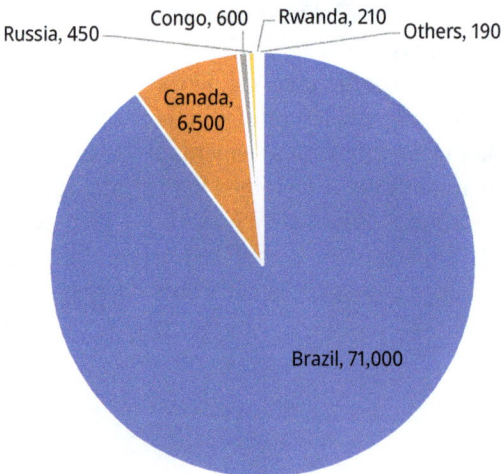

Figure 14.15: Niobium production, in metric tons.

14.10.2 Tantalum and niobium production

Both tantalum and niobium are co-produced from complex ores that also contain other metals [27]. Usually the other metal is the prime material that functions as the economic driver for the tantalum or niobium production, except in the case of coltan. In many cases, thorium or uranium is the main product to be extracted from the ore, and thus radioactivity protocols must be followed as the ore batch is refined. Uranium and thorium production are discussed in more detail in Chapter 16.

Coltan has been mined extensively in the central region of Africa, and has been in the news increasingly in the past decade because the slave labor conditions under which it has been mined, and because the smuggling and sale of it to concerns in the developed world has fed and ignited civil wars throughout the region, as well as the breakdown of government in the eastern Congo. The term "resource curse" was coined to describe this situation, in which a poor, underdeveloped country suddenly finds it has a great wealth of some commodity, and a lawless scramble ensues to attain and export that commodity. In 2012, a Bruges European Economic Policy Briefing discussed ways by which coltan export and the situation in eastern Congo could be brought under control, so that militias and civil wars could not be funded by its continued sale [28]. There have been several other calls throughout the world for regulation of its mining and extraction as well.

The reaction chemistry by which tantalum and niobium are extracted from ores can be complex, and depends on the amount of other minerals and metals in each ore batch, but simplified reaction chemistry can be presented. Both tantalum and niobium in the form of oxides are first separated from other metals in their ores. To an extent, this can be done through means already discussed, such as floatation or precipitation. The two target metals are then separated from each other by forming complexes, as shown in Scheme 14.11.

$$Ta_2O_5 + Nb_2O_5 + 24\ HF \longrightarrow 2\ H_2[TaF_7] + 2\ H_2[NbOF_5] + 8\ H_2O$$

Scheme 14.11: Tantalum and niobium complex formation.

The formation of these complexes is then followed by their extraction from an aqueous solution by cyclohexanone or other organic solvents. Scheme 14.12 shows how the niobium complex is then precipitated.

From this point, tantalum can be reduced to the metal using sodium at elevated temperatures (ca. 800 °C). While niobium can also be reduced, it is sometimes profitable to simply produce ferroniobium through a combination of its oxide with aluminum and iron(III) oxide in a type of thermite reaction or Goldschmidt reaction, the simplified reaction chemistry which is shown in Scheme 14.13.

$$2\ KF + H_2[NbOF_5] \longrightarrow 2\ HF + K_2[NbOF_5]_{(s)}$$

or

$$H_2[NbOF_5] + 10\ NH_4OH \longrightarrow 10\ NH_4F + 7\ H_2O + Nb_2O_{5(s)}$$

and

$$H_2[TaF_7] + 7\ NH_3 + 5\ H_2O \longrightarrow Ta(OH)_5 + 7\ NH_4F$$

Scheme 14.12: Tantalum and niobium complex precipitation.

$$3\ Nb_2O_5 + 12\ Al + Fe_2O_3 \longrightarrow 2\ Fe + 6\ Nb + 6\ Al_2O_3$$

Scheme 14.13: Ferroniobium production.

14.10.3 Tantalum and niobium uses

Concerning the uses of tantalum, the USGS states in its annual Mineral Commodity Summaries:

"Major end uses for tantalum included alloys for gas turbines used in the aerospace and oil and gas industries; tantalum capacitors for automotive electronics, mobile accessories, and personal computers; tantalum carbides for cutting and boring tools; and tantalum oxide (Ta_2O_5) was used in glass lenses to make lighter weight camera lenses that produce a brighter image" [2].

Indeed, it is the amazing rise in the production and use of cellular phones in the last two decades that has been linked to increasing coltan mining in the eastern Congo, and thus to the civil wars in that region. Because of this there have been increasing demands on the part of consumers and some governments to find ways to recycle or re-use the components of cellular phones, ultimately to lessen the demand for coltan.

Concerning the uses of niobium, the USGS Mineral Commodity Summaries state:

"Companies in the United States produced ferroniobium and niobium compounds, metal, and other alloys from imported niobium minerals, oxides, and ferroniobium. Niobium was consumed mostly in the form of ferroniobium by the steel industry and as niobium alloys and metal by the aerospace industry. Major end-use distribution of reported niobium consumption was as follows: steels, 75 %; and superalloys, about 25 %" [2].

Clearly, both metals are useful in several specific applications. The apparently ever-increasing use of both personal computers and cellular phones indicates that tantalum will continue to be needed. And one expects that the steel industry will continue to require a supply of niobium.

Figure 14.16: Cell phone recycling kiosk.

14.10.4 Tantalum and niobium recycling

Both these metals are distributed in small amounts into a large number of user end objects, and thus recycling of these elements becomes difficult in terms of economic viability. In the past few years, there have been an increasing number of kiosks in shopping malls throughout the world at which cell phones can be turned in for some amount of cash, as shown in Figure 14.16, a kiosk in a mall located north of Detroit, Michigan, USA. These represent a first attempt at the recycling of these metals and other rare earth metals that are used in cell phones.

14.11 Amalgams

The broad term "alloy" means any mixing of two metal elements (and is routinely expanded to include the nonmetallic element carbon, when discussing the production of steel). The term "amalgam" has been widely used by the general public, but in a chemical sense tends to mean any metal alloyed with mercury. Mercury and silver have in the past been used as dental amalgams, but have seen increased competition in recent years from ceramic composites. This is, in part, due to a perceived health risk related to mercury dental amalgams, even though many users have had such inserts for decades without health problems, and several studies have indicated that these dental amalgams are safe [29].

As mentioned in Section 14.4, what can be more loosely called gold-mercury amalgams continue to be used in the extraction of gold from gold-bearing sands. This remains economically practical when the size of the gold grains is small enough that they cannot be separated by any mechanical methods.

Bibliography

[1] Kroll Process. US Patent: 2.205.854, issued 25 June 1940.
[2] United States Geological Survey, Mineral Commodity Summaries, 2023. Website. (Accessed 18 December 2023 as: https://www.usgs.gov, https://doi.org/10.3133/mcs2023, as a downloadable pdf).
[3] ATI (Allegheny Technologies, Inc.). Website. (Accessed 22 December 2023, as: https://atimaterials.com).
[4] Strategic and Critical Materials 2013 Report on Stockpile Requirements, Office of the Undersecretary of Defense for Acquisition, Technology and Logistics. Website. Defense Logistics Agency, The Nation's Combat Logistics Support Agency. Website. (Accessed 22 December 2023, as: dla.mil/Strategic-Materials/Reports/).
[5] Goldschmidt, Dr. Hans; Vautin, Claude (1898-06-30). "Aluminium as a Heating and Reducing Agent". Journal of the Society of Chemical Industry 6 (17): 543–545.
[6] United States Department of Defense. Strategic and Critical Materials 2013 Report on Stockpile Requirements.
[7] World Gold Council. Website. (Accessed 22 December 2023, as: https://www.gold.org).
[8] The Silver Institute. Website. (Accessed 22 December 2023, as: https://www.silverinstitute.org).
[9] The Silver Institute News. Website. (Accessed 22 December 2023, as: https://www.silverinstitute.org/wp-content/uploads/2020/01/SNDec2019.pdf).
[10] The International Tin Association. Website. (Accessed 22 December 2023 as: https://www.internationaltin.org).
[11] EM Vinto. Website. (Accessed 22 December 2023 as: https://www.bnamericas.com).
[12] Gejiu Zi Li. Website. (Accessed 22 December 2023 as: https://www.mbdatabase.com).
[13] Huaxi Nonferrous Metals. Website. (Accessed 22 December 2023 as: https://china-tin.com).
[14] Malaysia Smelting Corp. Website. (Accessed 22 December 2023 as: https://www.msmelt.com).
[15] Aurubis. Website. (Accessed 22 December 2023, as: https://aurubis.com).
[16] Minsur. Website. (Accessed 22 December 2023 as: https://www.minsur.com).
[17] PT Timah. Website. (Accessed 22 December 2023 as: https://www.timah.com).
[18] Thiascaro. Thailand Smelting and Refining Company. Website. (Accessed 22 December 2023 as: https://www.thiascaro.com).
[19] Yunnan Cheng Feng, Ltd. Website. (Accessed 22 December 2023 as: www.internationaltin.org/tag/yunnan-chengfeng).
[20] Yunnan Tin Group. Website. (Accessed 22 December 2023 as: Web.archive.org/web/20090430133352/http://en.ytc.cn).
[21] Minerals Education Coalition. Website. (Accessed 22 December 2023, as: mineralseducationcoalition.org/elements/tin).
[22] Association of British Pewter Craftsmen, Website. (Accessed 22 December 2023, as: https://pewterers.org.uk).
[23] US Environmental Protection Agency. Website. (Accessed 22 December 2023 as: https://www.epa.gov/radiation/radionuclide-basics-technetium-99#).

[24] International Platinum Group Metals Association. Website. (Accessed 22 December 2023 as: https://ipa-news.com).

[25] Royal Society of Chemistry. Website. (Accessed 22 December 2023, as: https://www.rsc.org/periodic-table/element/43/technetium).

[26] United states Environmental Protection Agency. Website. (Accessed 22 December 2023, as: https://www.epa.gov/radiation/).

[27] Tantalum Niobium International Study Center. Website. (Accessed 22 December 2023, as: https://www.tanb.org).

[28] Bruges European Economic Policy Briefing, College of Europe. Website. (Accessed 22 December 2023, as: https://www.coleurope.eu/sites/default/files/research-paper/beep23_0.pdf, Coltan from Central Africa, International Trade and Implications for Any Certification).

[29] American Dental Association, amalgam. Website. (Accessed 22 December 2023, as: https://www.ada.org/en/resources/research/science-and-research-institute/oral-health-topics/amalgam).

15 Rare earth elements

15.1 Introduction

The lanthanide elements, also called the lanthanoids, as well as the older names the inner transition metals or the rare earth elements (often abbreviated REEs), are actually not as rare as the lattermost name implies. Several other elements that find large-scale uses are actually less common than some of the rare earths. But a common feature of the REEs is that they are widely distributed in small amounts among ores from which other elements are extracted, usually metals, and thus are not the primary material which is mined. Indeed, there are virtually no mining operations that produce a single REE as the primary or sole material.

A close examination of the names of the REEs reveals that several of them are in some way derived from the name Ytterby, Sweden. The names: yttrium, ytterbium, erbium, and terbium all have their root in the town name. This is because a mine near it contains these four elements, although their isolation from the ore named yttria and consequent discovery spanned almost a century. The elements gadolinium, thulium, and holmium were also found at this location.

The proper identification and discovery of several of the REEs took significant effort in the 19th century, with some of them not being purified to the metal until well into the 20th century. Table 15.1 shows a comparison of the REEs and several other elements in terms of their estimated abundance within the earth, and gives their year of discovery. Scandium and yttrium have also been included in the table because they are often comined with the REEs.

15.2 Isolation and production

Rare earth elements are often extracted and refined to their purified oxides, and this commodity is what is tracked by national governments and industry trade groups. Figure 15.1 shows the current production according to the United States Geological Survey's Mineral Commodity Summaries [1]. This is also tracked by the US Department of Defense in their Strategic and Critical Materials 2013 Report on Stockpile Requirements [2].

It is obvious from Figure 15.1 that China currently dominates the production of the REEs. This is, in part, because some mines in the United States and Canada have been closed since they are no longer economically profitable. Should the price of the REEs rise, these mines may again be worked. Despite only one operable mine in the United States, the United States Geological Survey has compiled reports indicating the locations of REE deposits within the United States, in the event that increased production is needed [3].

https://doi.org/10.1515/9783111329512-015

Table 15.1: Relative abundances of the REEs and selected other elements*.

Element	Symbol	Abundance (ppm)	Atomic number	Year of discovery
Zinc	**Zn**	**75**	**30**	Ancient
Cerium	Ce	68	58	1803
Copper	**Cu**	**51**	**29**	Ancient
Neodymium	Nd	33	60	1885
Lanthanum	La	32	57	1839
Yttrium	**Y**	**30**	**39**	1828
Cobalt	**Co**	**21**	**27**	1735
Scandium	**Sc**	**20**	**21**	1879
Lead	**Pb**	**20**	**82**	Ancient
Samarium	Sm	20	62	1879
Gadolinium	Gd	20	64	1886
Praseodymium	Pr	16	59	1885
Dysprosium	Dy	13	66	1950s
Ytterbium	Yb	10	70	1953
Hafnium	Hf	8	72	1923
Erbium	Er	7	68	1934
Tin	**Sn**	**3**	**50**	Ancient
Holmium	Ho	3	67	1878
Terbium	Tb	3	65	1843
Europium	Eu	2	63	1890
Lutetium	Lu	2	71	1907
Thulium	Tm	0.7	69	1911
Bromine	**Br**	**0.4**	**35**	1826
Uranium	**U**	**0.03**	**92**	1789

*NonREEs in bold

Table 15.2 provides a nonexhaustive, generally alphabetical list of minerals that have been found to contain REEs. Some ores are known by more than one name.

It can be seen from Table 15.2 that some of these minerals have very simple formulas, at least in their ideal sense. Fluorite – calcium fluoride or CaF_2 – is one example. As mentioned in Chapter 8, when fluorite is mined it can be clear or nearly clear, which could mean it forms naturally in a quite pure state. But even clear fluorite could have significant amounts of lanthanides within it, as these elements do not normally impart color to a mineral (the general exception to this rule being cerium).

Some of the minerals from which REEs are extracted are, on the other hand, quite complex in their formulas. Aeschynite, cerite, monazite, orthite, xenotime, and ytterbite are some obvious examples of this, although there are others. Ores that are this complex in their make-up require multiple chemical and physical processes to isolate single elements from them.

The reason there are so many different ores that contain one or more of the REEs is that their sizes and electronegativities are so close to each other that they become

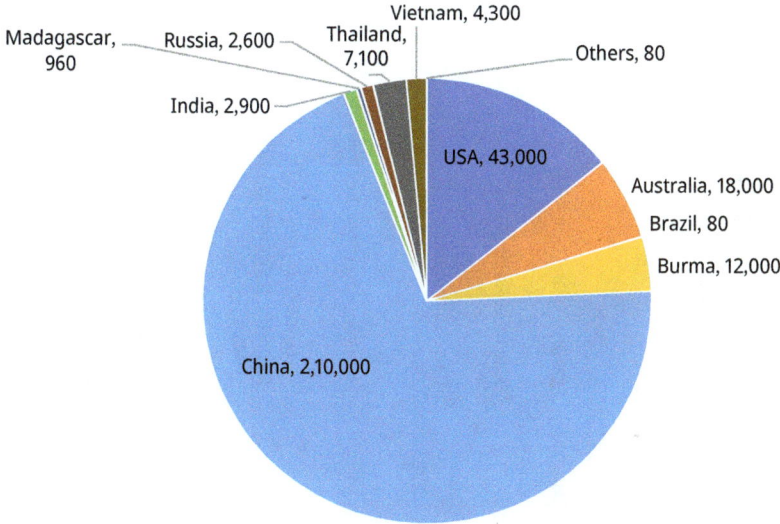

Figure 15.1: Rare earth oxide production.

essentially interchangeable in crystalline structures. These two properties not only ensure that they are distributed widely in numerous minerals, but also make the separation of one REE from another a difficult task that can consume significant amounts of time and energy [4]. The general techniques by which REEs are separated from each other and from other metals certainly have similarities to the purification and reduction steps used for other metals, but also have some characteristics that are unique to the REEs, and which have not been upgraded or changed for nearly a century.

15.3 Rare earth element purification

As mentioned, several of the REEs have names that are in some way connected to Ytterby, Sweden because the mine near there contains seven of them, and was the first from which these elements were discovered and separated. The isolation of the rare earth oxides and the isolation of the metals in some ways have not progressed noticeably from the first efforts that were formulated in Sweden. But all require some combination of different purification and reduction steps, and thus we will examine both.

Since there are no mines in which every REE is present, we should first examine what can be considered common steps in the refining and purification process.

1. *Milling and grinding.* Ores are first ground and crushed to uniform particle size, often that of grains of sand or even dust. The objective is to maximize surface area through this chemical process so that later chemical processes occur more quickly and uniformly.

Table 15.2: Rare earth element mineral sources.

Ore	General formula	Geographic location	Comments
Aeschynite	$(Nd, Ce, Ca, Th)(Ti, Nb)_2(O, OH)_6$	China, Inner Mongolia, USA	Sources can be high in Ce, Nd, or Y. REE within
Anatase	TiO_2	USA, France	Source of Ce or La
Ancylite	$Sr(Ce, La)(CO_3)_2(OH) \cdot H_2O$	USA	Can be found without rare earths.
Apatite	$Ca_5(PO_4)_3(F, Cl, OH)$	Apatity, Russia; Florida, USA; Canada	Significant source of Ce
Bastnasite	$(Ce, La, Y)CO_3F$	Sweden, Pakistan, USA	Source of U
Brannerite	$(Ca, Y, Ce, U)(Ti, Fe)_2O_6$	USA	
Britholite	$(Ce, Ca)_5(SiO_4, PO_4)_3(OH, F)$	USA	
Brockite	$(Ca, Th, Ce)PO_4 \cdot H_2O$	Colorado, USA	
Cerianite	$(Ce, Th)O_2$	USA	As Ce^{4+}
Cerite	$(Ce, Ca, La)_9(Mg, Fe^{3+})(SiO_4)_6(SiO_3OH)(OH)_3$	Vastmanland, Sweden; Mountain Pass, California, USA; Kola, Russia	Cerite-(Ce) and cerite-(La)
Cheralite	$(Ca, Ce, Th)(P, Si)O_4$	USA	Also recoverable Th
Chevkinite	$(Ca, Ce, Th)_4(Fe, Mg)_2(Ti, Fe)_3Si_4O_{22}$	USA	
Churchite	YPO_4H_2O	USA	Can also contain REEs
Crandallite	$CaAl_3(PO_4)_2(OH)_5 \cdot H_2O$	USA	
Doverite	$YcaF(CO_3)_2$	USA	
Eudialyte	$Na_4(Ca, Ce)_2(Fe, Mn, Y)ZrSi_8O_{22}(OH, Cl)_2$	USA	Also contains, U, Nb, Ta, Hf
Euxenite	$(Y, Ca, Ce, U, Th)(Nb, Ta, Ti)_2O_6$	USA, Norway	
Fergusonite	$YNbO_4$	USA	Fergusonite-(Y) and -(Ce)
Fluocerite	$(La, Ce)F_3$	Sweden; Kazakhstan; Australia: Inner Mongolia, China	Fluocerite-(La) and fluocerite-(Ce)
Fluorapatite	$(Ca, Ce)_5(PO_4)_3F$	USA	
Fluorite	CaF_2	Very widespread	Y, Yb, and Eu in fluorite often account for the fluorescence
Gagarinite	$NaCaY(Cl, F)_6$	USA	
Gerenite	$(Na, Ca)_2Y_3Si_6O_{18} \cdot 2H_2O$	USA	REEs as well as Y
Gorceixite	$BaAl_3[(PO_4)_2(OH)_5] \cdot H_2O$	USA	
Goyazite	$SrAl_3(PO_4)_2(OH)_5 \cdot H_2O$	USA, France	

Table 15.2 (continued)

Ore	General formula	Geographic location	Comments
Hingganite	$(Y, Yb, Er)_2Be_2Si_2O_8(OH)_2$	USA	
Ilmorite	$Y_2(SiO_4)(CO_3)$	USA	
Kainosite	$Ca_2(Y, Ce)_2Si_4O_{12}(CO_3) \cdot H_2O$	USA, Norway	
Loparite	$(Ce, Na, Ca)(Ti, Nb)O_3$	USA Russia	
Monazite	$(La, Ce, Pr, Nd, Y, Th)PO_4$	India; Madagascar; South Africa; Bolivia; Australia	Four different types
Orthite (aka. Allanite)	$(Ce, Ca, Y, La)_2(Al, Fe^{3+})_3(SiO_4)_3(OH)$	Greenland; Queensland, Australia; New Mexico, USA	Designated allanite-(Ce), allanite-(La) or allanite-(Y).
Parasite	$Ca(La, Ce)_2(CO_3)_3F_2$	Colombia; Greenland	Can contain Nd
Perovskite	$CaTiO_3$	USA, Russia, Sweden	May contain Nb
Pyrochlore	$(Na, Ca)_2Nb_2O_6(OH, F)$	Norway	May contain REE and transition metals
Rhabdophane	$(Ce, La)PO_4 \cdot H_2O$	USA	Rhabdophane-(Ce) and -(La), may contain Nd
Rinkite	$(Ca, Ce)_4Na(Na, Ca)_2Ti(Si_2O_7)_2F_2(O, F)_2$	Russia, Greenland	
Samarskite	$(YFe^{3+}Fe^{2+}U, Th, Ca)_2(Nb, Ta)_2O_8$	USA, Russia	
Stillwellite	$(Ca, Ce, La)BSiO_5$	Queensland, Australia; Tajikstan; Ontario, Canada	
Synchysite	$Ca(Ce, La)(CO_3)F$	USA	Synchysite-(Ce) and -(Y) and -(Nd)
Thalenite	$Y_3Si_3O_{10}(F, OH)$	USA	Can occur in zircon
Thorite	$(Th, U)SiO_4$	Norway, USA	
Titanite	$CaTiSiO_5$	Very widespread	Fe, Al, Ce, Y, and Th can be present
Uraninite	$(U, Th, Ce)O_2$	USA, Germany	Usually mined for Th
Vitusite	$Na_3(Ce, La, Nd)(PO_4)_2$	USA	Vitusite-(Ce)
Wakefieldite	$(L, Ce, Nd, Y)VO_4$	Canada; Congo	Four types, based on dominant rare earth
Xenotime	$(Y, Yb, Dy, Er, Tb, U, Th)PO_4$	Brazil; Norway	
Ytterbite (aka. Gadolinite)	$(La, Ce, Nd, Y)_2FeBe_2Si_2O_{10}$	Norway; Sweden; Colorado, USA	Gadolinite-(Y) or gadolinite-(Ce)
Yttrofluorite	$(Ca, Y)F_2$	USA, Sweden	May also contain Tb
Zircon	$ZrSiO_4$	Australia	May contain traces of Hf, U, Th.

2. *Electromagnetic separation.* This broad step takes advantage of the fact that some ores such as bastnäsite and monazite are magnetic, and thus can be separated from nonmagnetic material using a conveyor belt with magnets in one of the end rollers. Both magnetic and nonmagnetic ores will drop from the belt as it rolls over, but nonmagnetic materials will drop off first, with magnetic material remaining on the roller until the belt and magnetic roller are separated. This step may need to be performed many times to ensure full separation.

3. *Ore flotation.* Froth flotation was discussed in Chapter 13, since it had been patented for the concentration of copper ores. It is beneficial in the concentration of REEs as well, using a stream of air injected into an aqueous solution to separate the desired ore from impurities which settle out of solution. This step may also need to be performed repeatedly.

4. *Centrifugal concentration.* The use of high-speed centrifuges furthers the separation of materials based on small differences in densities of the particles. Once again, this step represents a process that may need to be repeated many times.

5. *Leaching and precipitation.* This broad step can be further broken up into parts, including the following:

    (a) *Fractional crystallization.* Materials are dissolved in a liquid, often water, then through careful control of temperature and pH, selective precipitation of less soluble materials takes place.

    (b) Ion exchange. A solution containing the REEs is filtered through some medium, such as a resin or zeolite, where the REE ions are bound to the material. These materials are then washed repeatedly with various solvents, extracting different elements for each specific solvent. As with other steps, this may need to be performed numerous times.

    (c) Extraction via solvents. The principle applied in this step is to dissolve a mixture of two REEs in two solvents which are each immiscible in the other. The rare earths should have slightly differing solubilities in the two solvent liquids, and thus each will concentrate in a different liquid. This can also involve the use of various acid or base solutions to precipitate out one or more elements selectively. For example, thorium precipitates out of monazite ores in highly basic solutions, usually as a phosphate, leaving REEs in solution for further separation.

6. Electrolytic deposition. When metal salts of the REEs are soluble in various solutions, they can be electrodeposited. This usually requires a precise control of the electrical flow, as well as some anodic material that can be made soluble in the cell.

15.3.1 Yttrium isolation

Yttrium occurs in several of ores that also contain REEs, and thus yttrium is at times co-produced with them. Even though it is lighter than all of the REEs, its size and reactivity are more closely aligned with the heavy REEs and thus it tends to be concentrated with the heavy REEs as they are being refined.

Yttrium continues to be a niche use metal, and the amounts required remain small, certainly compared to some other industrially useful metals. Yttrium in ceramics is in the form of yttria, Y_2O_3, which is usually mixed with zirconia as a stabilizer. Likewise, its major use in phosphors is as yttria or in some doped form of yttria.

When yttrium metal must be refined, the traditional method utilized crucibles made from either tungsten or tantalum. In 1979, a US Patent was filed for the production of yttrium in a less expensive manner, using an iron crucible with a tap hole, and calcium fluoride to coat the crucible. Then yttrium fluoride and calcium metal were electrolyzed resulting in the reduced yttrium metal. The simplified reaction is shown in Scheme 15.1.

$$2\,YF_{3(l)} + 3\,Ca_{(l)} \longrightarrow 3\,CaF_{2(l)} + 2\,Y_{(l)}$$

Scheme 15.1: Yttrium metal reduction.

In practice, the two materials are not added in a 2:3 ratio; rather a 12:5 ratio by weight has proven to be most effective.

15.3.2 Isolation of rare earth elemental metals

Each of the isolated rare earth element oxides can be reduced, once they have been concentrated, usually using a combination of acid or base digestion, followed by an electrolytic process such as electrowinning [4].

Current processes for the reduction of the heavy REEs remain small-scale processes, as there are no large-scale uses for the last five elements in this series.

15.4 Rare earth element uses

Numerous uses for the REEs have developed in the past few decades. Some are exclusive to only one or two elements, while other applications utilize numerous different elements [3]. Table 15.3 gives a general division of where each of these elements is utilized. Note that the heavy REEs – holmium, erbium, thulium, ytterbium, and lutetium – currently have no major industrial applications and uses. Additionally, promethium is not listed because it is radioactive and all its isotopes have very short half-lives. Yttrium has also been included, because it is co-produced with the lighter REEs.

Table 15.3: Rare earth element uses*.

	Magnets	Batteries	Metal alloys	Automotive catalysts	Petroleum refining	Polishing materials	Glass mixes	Phosphors	Ceramics
Y							2	69.2	53
La		50	26	5	90	31.5	24	8.5	17
Ce		33.4	52	90	10	65	66	11	12
Pr	23.4	3.3	5.5	2		3.5	1		6
Nd	69.4	10	16.5	3			3		12
Sm		3.3							
Eu								4.9	
Gd	2							1.8	
Tb	0.2							4.6	
Dy	5								

*The percentages for each column may not add up to 100 because of the possibility for several other, smaller niche uses for that REE either as an oxide or as the elemental metal.

15.4.1 Alloys

As can be seen in Table 15.3, cerium finds use in a variety of alloys, usually aluminum alloys as well as iron alloys. As well, lanthanum is used in alloys, usually as an additive in iron alloys. More recently, lanthanum has been used in some hydrogen sponge alloys.

The use of neodymium in alloys is essentially that of neodymium–iron–boron alloys that are used as magnets, which is the next topic of discussion.

15.4.2 Magnets

It is apparent from the Table 15.3 that neodymium finds its largest use in magnets, although it has other applications as well. The iron-neodymium-boron magnets, sometimes called NIB or Neo magnets, have the formula $Nd_2Fe_{14}B$, are the strongest known by weight, and now are used in a wide variety of applications, having even been used in the craft that NASA has sent to Mars. More domestically, they find use in hard drives, cordless power tools, and other applications where the weight of the consumer end object is of concern, and needs to be as low as possible. The technology for their production has become mature and widespread, having been first discovered in the 1980s.

Praseodymium also finds use in magnets, as a component of neodymium magnets. The end result is again high power, high-strength magnets that can be used in applications where small size is required.

Dysprosium can also be substituted into NIB magnets, but only to 6 % of the total neodymium. This makes the resulting magnet more resistant to corrosion, and has been examined, but not yet adopted, for use in electric automobile motors.

Additionally, magnets that include one or more REEs have found use in large, modern wind turbines [5].

15.4.3 Batteries

What are called nickel-metal-hydride batteries routinely have lanthanum in them as the "metal." The batteries used in new electric automobiles such as the Prius routinely require over 20 pounds of lanthanum (9 kg) and over 2 pounds of neodymium (0.9 kg). As the number of such vehicles is sold, the amount of each of these REEs that is available may conceivably become a bottleneck to production.

In the last 10 years, cerium has found use in what is now called a zinc-cerium battery. This couple sets up a large voltage (2.43 V), and is rechargeable. It is also somewhat less expensive than comparable vanadium batteries, and may thus find a larger market in the future.

15.4.4 Catalysts

A significant amount of cerium is used in automotive catalysts in the form of ceria. While platinum or palladium is considered to be the catalyst in the catalytic convertor of an automobile, cerium is in what is called the "washcoat" of the catalyst, enhancing the surface area of the catalyst so that it can function more efficiently, and lowering the amount of the more expensive platinum which is used.

15.4.5 Petroleum refining

Lanthanum finds use in the petroleum refining process in steps that involve the use of zeolites. Lanthanum-containing zeolites tend to be stable at high temperatures, and thus perform well at the temperatures required for petroleum cracking. Cerium is incorporated in similar fashion.

15.4.6 Polishing materials

Ceria is the oxidized form of cerium that is used extensively in polishing materials. In many cases, a slurry is made of ceria and some solvent, and then used to polish surfaces, often glass surfaces, such as high-quality optical lenses. This use has become widespread, and there is enough overall need that the USGS Mineral Commodity Summaries delineates cerium in its annual assessment of the REEs [1].

15.4.7 Glasses and glass mixes

Table 15.3 indicates that several of the REEs are used in some form of glass mix. Cerium oxide is used in glass formulations in televisions to prevent darkening of the screen. Lanthanum is required in gallium lanthanum sulfide glass, which has several small, niche uses.

15.4.8 Ceramics

Yttrium finds use in ceramics in what is often abbreviated YSZ, which stands for yttria-stabilized zirconia. It is the zirconia – more properly, zirconium dioxide – that is the major component of this type of ceramic. But the addition of what are usually proprietary amounts of yttria, or Y_2O_3, stabilizes the end product. These specialized ceramics have a number of small-scale, niche uses. One of the most interesting is arguably non-metal knife blades. This has caused some problems in the recent past as such blades are not detected in metal detectors.

15.5 Recycling

The recycling of REEs has not yet become a widespread part of the metals recycling programs in most developed countries. There have in the past 5 years been some programs put into place in malls in the United States for the recycling of cell phones, usually for some cash back to the person turning in the phone. Such programs are aimed at recycling all the components of the phone, which includes the neodymium-containing magnet [6–8].

Bibliography

[1] United States Geological Survey, Mineral Commodity Summaries, 2023. Website. (Accessed 18 December 2023 as: https://www.usgs.gov, https://doi.org/10.3133/mcs2023, as a downloadable pdf).
[2] Department of Defense. Strategic and Critical Materials 2013 Report on Stockpile Requirements. Website. (Accessed 22 December 2023, as: https://www.mineralsmakelife.org/assets/images/content/resources/Strategic_and_Critical_Materials_2013_Report_on_Stockpile_Requirements.pdf).
[3] USGS, The Principal Rare Earth Elements Deposits of the United States – A Summary of Domestic Deposits and a Global Perspective. Website. (Accessed 23 December 2023, as: https:pubs.usgs.gov/sir/2010/5220).
[4] Holleman and Wiberg. *Inorganic Chemistry*, 1995, DeGruyter.
[5] Mining Watch Canada. Website. https://www.miningwatch.ca/blog/2013/5/17/rare-earth-elements-background-information).
[6] Resources for the Future. RFF Report. The Supply Chain and Industrial Organization of Rare Earth Materials: Implications for the U.S.Wind Energy Sector. Website. (Accessed 23 December 2023, as:

https://www.rff.org/publications/reports/the-supply-chain-and-industrial-organization-of-the-rare-earth-materials-implications-for-the-us-wind-energy-sector/).

[7] Critical Minerals Centre of Excellence. Website. (Accessed 23 December 2023, as: Canada.ca/en/campaign/critical-minerals-in-canada.html).

[8] Handbook of Rare Earth Elements. Website. (Accessed 23 December 2023, as: https://www.degruyter.com/document/doi/10.1515/9783110365085/html?lang=en).

16 Uranium and thorium

16.1 Introduction

Uranium and thorium are both elements with relatively short histories and few uses. The discovery of these elements occurred as the seventeenth century turned to the eighteenth, but any large-scale uses for them did not come about until the twentieth. Interestingly, their study involved such famous names as Marie and Pierre Curie, who first isolated the element, and Henri Becquerel, who first realized the penetrating power of the radiation which is emitted from it. Becquerel's famous and most likely inadvertent experiment, in which he exposed an undeveloped photographic plate to a uranium salt, has become something of a classic example that teachers use even today to admonish students to keep an open mind (as Becquerel did when investigating the phenomenon). Becquerel and the Curies shared the 1903 Nobel Prize in Physics for their work in the area of radioactivity.

16.2 Sources

Several of the ores listed in Table 15.1 in Chapter 15 discussing rare earth elements also contain either uranium or thorium. When uranium is the main element to be refined from a particular ore, there are several that can be mined profitably. Canada, Australia,

Table 16.1: Uranium-bearing ores.

Name	Formula*	% Uranium	Location
Autunite	$Ca(UO_2)_2(PO_4)_2 \cdot 8 H_2O$	52.1	France, USA
Brannerite	UTi_2O_6	55.3	
Carnotite	$K_2(UO_2)_2(VO_4)_2 \cdot 3 H_2O$	52.8	USA, Congo, Morocco, Australia, Kazakhstan
Coffinite	$USiO_4(OH)_4$	80.4	USA
Davidite	$(Ln)(Y, U)(Ti, Fe^{3+})_{20}O_{38}$**		Australia, Norway
Saleeite	$Mg(UO_2)_2(PO_4)_2 \cdot 10 H_2O$	51.0	Congo
Thucholite	Mix of uraninite, hydrocarbons and sulfides		
Torbernite	$Cu(UO_2)_2(PO_4)_2 \cdot 12 H_2O$	47.1	
Tyuyamunite	$Ca(UO_2)_2(VO_4)_2 \cdot 8 H_2O$	50.0	Kyrgystan (rare)
Uraninite (aka pitchblende)	UO_2	88.1	Congo, Canada, Australia
Uranocircite	$Ba(UO_2)_2(PO_4)_2 \cdot 10 H_2O$	45.5	Germany
Uranophane	$Ca(UO_2)_2(HSiO_4)_2 \cdot 5 H_2O$	55.6	
Zeunerite	$Cu(UO_2)_2(AsO_4)_2 \cdot 10 H_2O$	44.8	Germany

*formulas are given in their simplest form, **Ln = some lanthanide

https://doi.org/10.1515/9783111329512-016

and Kazakhstan all have large deposits of uranium ores, although several other countries have deposits of one uranium ore or another. The United States Geological Survey Mineral Commodity Summaries does not track uranium production, although it does do so for thorium [1]. The United States Environmental Protection Agency (EPA) does disseminate information on what protocols must be undertaken when working with uranium and thorium [2]. Table 16.1 lists uranium ores alphabetically.

Clearly, urananite or pitchblende has the highest percentage of uranium in it. But since almost all ores exist with other materials, usually silicates, mixed within them, each ore is examined for its starting mineral concentration. At present, the uraninite deposits in Canada have proven to be the most profitable to exploit, as their content can be as high as 20 % uranium. Table 16.2 shows the top 10 uranium producing mines in the world, where it can be seen that Canadian mining operations dominate [3–13].

Table 16.2: Top 10 uranium producing mines world-wide.

Name	Country	Corporation	Amount (tons U)
McArthur River	Canada	Cameco	7,686
Olympic Dam	Australia	BHP Billiton	3,353
Arlit	Niger	Somair/Areva	2,726
Tortkuduk	Kazakhstan	Katco JV/Areva	2,608
Ranger	Australia	Rio Tinto	2,240
Kraznokamensk	Russia	ARMZ	2,191
Budenovskoye 2	Kazakhstan	Karatau JV/Kazatomprom	2,175
Rossing	Namibia	Rio Tinto	1,822
Inkai	Kazakhstan	Inkai JV/Cameco	1,602
South Inkai	Kazakhstan	Betpak Dala JV	1,548

Monazite ore has proven to be profitable to mine in terms of the rare earths, and then in terms of thorium as a second product. Concerning thorium, the United States Geological Survey reports in its annual Mineral Commodity Summaries that:

"The world's leading thorium resources are found in placer, carbonatite, and vein-type deposits. Thorium is found in several minerals, including monazite, thorite, and thorianite. According to the World Nuclear Association, worldwide identified thorium resources were an estimated 6.4 million tons of thorium. Thorium resources are found throughout the world, most notably in Australia, Brazil, India, and the United States" [1].

Perhaps obviously, thorium production is connected with that of the lanthanides, or rare earth elements. But also, this official statement indicates that global and domestic demand for thorium is not particularly high. In Section 16.4, below, we discuss the uses and potential uses of thorium in more detail. Figure 16.1 shows the global breakdown of thorium reserves, in metric tons of ThO_2 often called thoria.

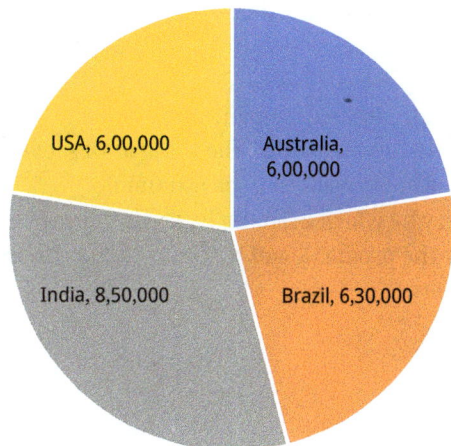

Figure 16.1: Thorium worldwide reserves.

16.3 Purification

Unlike metals such as gold, silver, copper, and iron, both uranium and thorium have relatively short histories, from their discovery in the late nineteenth century to their isolation and use now. Uranium's first discovery claim dates back to 1789. Thorium was first discovered in 1828 by Jons Jakob Berzelius, but the elements remained something of curiosities for decades. The term 'radioactivity' was coined by Marie Curie, who found that both elements were indeed radioactive.

16.3.1 Uranium purification

Uranium ores usually contain uranium oxides. The World Nuclear Association provides information concerning their reduction and ultimate use [2]. But, as with many materials, uranium goes through several preliminary steps in its concentration and refining. They include:

1. Crushing the ore batch. This is done to homogenize particle size, and routinely results in material that is the consistency of dust.
2. Dissolving the powdered ore in an acid. Nitric acid can be used to produce $UO_2(NO_3)_2 \cdot 6\,H_2O$, a solution of uranyl nitrate.
3. Precipitation and extraction, often using tributyl phosphate in kerosene. This concentrates the uranium in what is called the organic extractant, but may need to be repeated several times.
4. The concentrated uranium is then heated strongly to produce UO_3, or UO_2 if heated strongly enough.

The aim of this purification process is often not the production of reduced uranium metal, but rather the production of UF_6 gas. This can be isotopically enriched for use as a nuclear power source, or if enriched sufficiently, for weaponry. The reason a fluoride-containing uranium material has proven best for later enrichment is that fluorine has only one isotope, 19-F. This means that the separation based on the mass of UF_6, even though it is small, is based solely on different isotopic masses of the uranium.

Scheme 16.1 shows the simplified chemistry by which uranium hexafluoride is produced. A reducing atmosphere is required for the production of uranium dioxide, then hydrofluoric acid is required for the gasification.

$$UO_3 + H_{2(g)} \longrightarrow UO_2 + H_2O_{(g)}$$

or

$$U_3O_8 + 2H_{2(g)} \longrightarrow 3UO_2 + 2H_2O_{(g)}$$

followed by

$$UO_2 + 4HF_{(g)} \longrightarrow UF_{4(g)} + 2H_2O_{(g)}$$

followed by

$$UF_{4(g)} + F_{2(g)} \longrightarrow UF_{6(g)}$$

Scheme 16.1: UF_6 production.

The final step is run at high temperature, and must be carefully controlled because of the presence of elemental fluorine. The uranium hexafluoride can be stored, although care must be taken in its handling, since it is very corrosive. At elevated pressure, UF_6 can be liquefied, which makes handling somewhat easier.

16.3.2 Thorium purification

Again, thorium is extracted from monazite ores, almost always in conjunction with rare earth elements (REEs). Acid digestions are usually used to effect the initial separation of thorium from the REEs, because of their differing solubilities. Scheme 16.2 shows the simplified chemistry for this.

$$H_2SO_4 + LnO_x \longrightarrow LnSO_{4(aq)}$$

followed by

$$LnSO_{4(aq)} + NaOH \longrightarrow Th(OH)_{4(s)} + LnSO_{4(aq)}$$

Scheme 16.2: Thorium separation from the REEs.

The symbol Ln can represent any of several of the lanthanides, and is used so because each ore batch is initially different. This is not the only method whereby thorium can be separated from the REEs in an ore. Again, depending on the starting ore, nitric or hydrochloric acids are used for the initial digestion, and organic solvents can be used to remove thorium, usually as a hydroxide.

Elemental thorium is usually reduced from an oxide by the use of a reducing agent such as another metal. Calcium has been used successfully in this capacity. Scheme 16.3 shows the simplified chemistry for this:

$$2Ca + ThO_2 \longrightarrow Th + 2CaO$$

Scheme 16.3: Thorium reduction to the metal.

Elemental calcium can also be used as a reducing agent in the production of zirconium or uranium to their reduced forms. This becomes a major industrial use of calcium metal, as there are very few other reasons that calcium compounds are reduced to the metal.

16.4 Uranium and thorium uses

16.4.1 Uranium uses

Uranium has been used both for nuclear power plants as well for the fuel of nuclear weapons, but it has some other more minor uses as well. Small amounts of uranium have been used in different forms of glass and in different glazes.

Uranium was a popular additive to glass prior to the Second World War, and concentrations as low as 2 % impart a yellow to greenish color. The intensity of the color increases as the concentration of uranium – routinely in a diuranate ion (UO_4^{2-}) – increases in the glass formulation. Uranium glass is no longer made in large quantities.

Uranium glazes were used in tile ware prior to the Second World War as well, and can impart a wide range of colors, from pale yellows, through greens and blues.

16.4.2 Thorium uses

Thorium has enormous potential to be used as a nuclear power source, but thus far has not been utilized in this capacity. All existing, commercially productive nuclear power plants function with uranium fuels; and this is actually based only on the development of uranium as the fuel for an atomic weapon through the Manhattan Project in World War II. This use may change in the near future, as monazite deposits in India are large enough that thorium may be extracted profitably to make a thorium-based reactor. A recently published "Thorium Energy Report" discusses in considerable detail the progress that has been made in India recently towards constructing a fully functional nuclear power plant which runs on thorium-based fuel [4]. This holds enormous promise, since the waste products from such a power plant are much more difficult to refine into a material or materials that can be used for weaponry.

Thorium does have some other uses that have been exploited in the past, a list of which would include:

- Lantern mantles: thoria on the mantle imparts a very bright white light.
- Ceramics: thoria can be used to produce highly heat resistant ceramic materials for niche uses.
- Welding rods: small amounts of thorium as an additive create better burning rods
- Ophthalmic lenses: ThF_4 can be used as an anti-reflecting material on certain lenses. ThF_4 is produced by the direct combination of thorium with fluorine gas.
- Alloys for the aerospace industry: what is sometimes called Mag-Thor are alloys of magnesium, thorium, and zirconium, used in aerospace engine parts because of the material's high strength and light weight. Even with thorium concentrations at 1–2 % however, there is concern about radioactivity in the alloy, and the care required for its disposal or re-use.

16.5 By-products

The by-products of uranium refining are much the same as that found in other mining operations, namely what are called overburden waste and tailings. These are essentially all the rock and solid matter that does not contain the uranium. They can be radioactive, largely because uranium often occurs with some amount of radium or radon. After a mining operation ceases, this waste is often covered with earth as a form of landscaping, and also to prevent contamination of a wider area with gamma radiation [3].

As well, the refining of uranium ores to UF_6 requires significant amounts of fluorine, either as the element, or as hydrofluoric acid. These materials must be treated with care because of the corrosive nature of each. When economically feasible, fluorine-containing materials are re-used.

Spent uranium fuel rods are accounted for by nuclear power plants. They are often stored under water to moderate their radioactivity to the surroundings. Such spent nuclear fuel can be re-processed to extract specific elements and isotopes for further uses.

The by-products of thorium refining usually include the solvents and other materials by which rare earth elements are extracted and separated from the thorium. This can add up to a significant amount of material overall, since the extraction steps for rare earth elements are seldom quick, one-step processes. Indeed, rare earth element separation, including the separation of thorium, remains a very inefficient, expensive process.

16.6 Recycling and re-use

Thus far, uranium has been re-used in the form of re-refining spent fuel rods to obtain other elements, such as plutonium and technetium. Beyond this, governments track the use of uranium, but there are no recycling programs.

There are no recycling programs in what has become the traditional sense for any thorium-containing products [14].

Bibliography

[1]   United States Geological Survey, Mineral Commodity Summaries, 2023. Website. (Accessed 18 December 2023 as: https://www.usgs.gov, https://doi.org/10.3133/mcs2023, as a downloadable pdf).
[2]   Environmental Protection Agency. Website. (Accessed 23 December 2023, as: https://www.epa.gov/radiation/radionuclide-basics-thorium).
[3]   World Nuclear Association. Website. (Accessed 23 December 2023, as: https://world-nuclear.org/information-library/nuclear-fuel-cycle/introduction/nuclear-fuel-cycle-overview.aspx).
[4]   Cameco. Website. (Accessed 23 December 2023 as: https://www.cameco.com/businesses/uranium-operations/canada/mcarthur-river-key-lake).
[5]   Mining Technology.com Website. (Accessed 23 December 2023 as: https://www.mining-technology.com/).
[6]   Areva in Niger. Website. (Accessed 23 December 2023 as: https://www.orano.group/fr?).
[7]   Orano Group. Areva. Website. (Accessed 23 December 2023 as: https://www.sa.areva.com/news-kazakhstan-areva-and-kazakhstan-sign-a-strategic-agreement).
[8]   ERA, Energy Resources of Australia. Website. (Accessed 23 December 2023 as: http://www.energyres.com.au/).
[9]   ARMZ Uranium Holding Company. Website. (Accessed 23 December 2023 as: https://www.armz.ru).
[10] Uranium One Group, Rosatom. Website. (Accessed 23 December 2023 as: uranium1.com).
[11] Rio Tinto. Rossing Uranium. Website. (Accessed 21 April 2015 as: http://www.rossing.com/). Sold to China National Uranium Corporation in 2019.
[12] Cameco. Inkai. Website. (Accessed 23 December 2023 as: https://www.cameco.com/businesses/uranium-operations/kazakhstan/inkai).
[13] Uranium One, South Inkai. Website. (Accessed 23 December 2023 as: https://www.uranium1.com/index.php/en/mining-operations/kazakhstan/south-inkai-mine).
[14] IAEA International Atomic Energy Agency. Website. (Accessed 23 December 2023, as: www.iaea.org/newscenter/news/thoriums-long-term-potential-in-nuclear-energy-new-iaea-analysis).

17 Silicon

17.1 Introduction

Silicon is considered by some to be a metalloid, and by others to be a nonmetal. It never occurs in its reduced form in nature, yet when reduced, has a silvery, metallic appearance. Perhaps ironically for those who categorize it as a metalloid or nonmetallic element, its largest use is as a metal in the steel industry. The United States Geological Survey (USGS) Mineral Commodity Summaries routinely refer to silicon as a metal [1].

17.2 Sources

The USGS does track the worldwide production of silicon in its annual Mineral Commodity Summaries [1]. This is mostly because a correlation can be made between silicon production and ferrosilicon steel alloy production. Figure 17.1 shows the current global output in thousands of metric tons.

The dominance of China is obvious in Figure 17.1. But, as mentioned, the production of silicon goes along with the production of iron and steel, which were shown in similar fashion in Chapter 11. In those two areas as well, China is currently the world's largest producer.

There are several major corporations that produce silicon, and that do so for a variety of reasons. Table 17.1 is a nonexhaustive compilation of them.

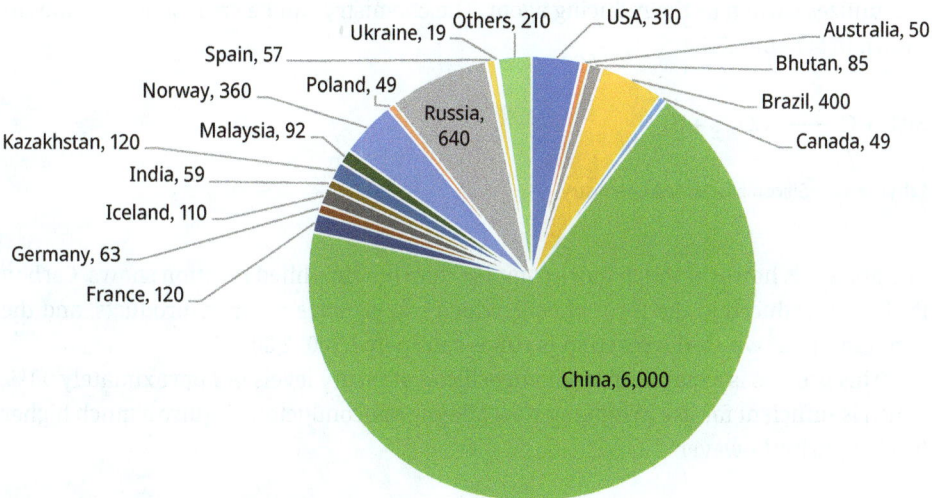

Figure 17.1: Silicon production.

https://doi.org/10.1515/9783111329512-017

Table 17.1: Silicon producers.

Company name	Silicon grade	Other products	Comments	Ref
China National Bluestar (Group) Co., Ltd.		Rubber, silica		[2]
Elkem	Solar silicon	Carbon and microsilica	Produces both high-grade silicon and ferrosilicon	[3]
Globe Specialty Metals	High purity	Silicones		[4]
Grupo Ferroatlántica	Solar silicon	Mn- and Fe–silicon alloys, energy		[5]
Hankook Silicon		Polysilicon		[6]
JFE Steel	Si–steel alloys	Steel alloys, titanium		[7]
Renewable Energy Corp.	High purity	Solar panels		[8]
Simcala, Inc.	High purity		Purchased by Dow	[9]
Wacker Chemie AG		Silicones, silanes		[10]

17.3 Purification

17.3.1 Metallurgical grade silicon

Silica is one of the most common lithic materials on the planet, although much of it is not pure, being mixed with other elements in their oxide or sulfide forms, such as iron, aluminum, or copper. Silicon production starts with a clean source of silicon dioxide (SiO_2), and utilizes carbon as the reducing agent. The chemistry can be represented simply, as shown in Scheme 17.1.

$$SiO_2 + C \longrightarrow CO_{2(g)} + Si$$

Scheme 17.1: Silicon production chemistry.

The process is however much more complex than the simplified reaction shows. Carbon is often introduced in the form of coal, which can produce other by-products, and the temperature at which the reaction is run is routinely 1,500–2,000 °C.

This process is capable of producing silicon at purity levels of approximately 98 %, which is sufficient for use in making steel alloys. Semiconductors require a much higher level of purity, however.

17.3.2 Semiconductor grade silicon

The production of semiconductor grade silicon is a much smaller operation than the production of silicon for ferrosilicon alloys, but it is one upon which many industries now depend. The reaction chemistry for this is shown in Scheme 17.2.

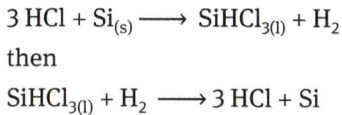

$$3\ HCl + Si_{(s)} \longrightarrow SiHCl_{3(l)} + H_2$$
then
$$SiHCl_{3(l)} + H_2 \longrightarrow 3\ HCl + Si$$

Scheme 17.2: Production of high-purity silicon.

The silicon must be in powdered form for this reaction, and the working temperature is routinely 300 °C. Any impurities are also converted into halides, and the silicon halide is separated from those of the impurities through distillation, because the silicon halide has a very low boiling point, 32 °C. The final step, the reduction back to elemental silicon, uses hydrogen as the reducing agent, and requires 8–13 days at 1,100 °C.

This process is still called the Siemens process in some literature, because it was developed and pioneered in the Siemens' labs over 50 years ago. It functions by a chemical vapor deposition of silicon onto existing high-purity silicon rods at the above-mentioned temperatures. This ultimately can bring impurities in the silicon down to the parts per billion levels.

More recently, a fluidized bed reactor process has been established that utilizes silane (SiH_4) as a starting material, which is injected into the reaction vessel as seed particles of pure silicon are introduced from the top of the chamber. As the silane reacts to form silicon on the seed crystals, they grow large enough to fall to the bottom of the chamber where they are recovered. This produces extremely high-purity silicon. This technology remains however a less common technique than the more established Siemens process.

17.3.3 Single crystal high-purity silicon

The production of single crystals of high-grade silicon can be accomplished by what is known as the Czochralski process. This process can be used to manufacture single crystals of metals as well, but currently the production of silicon crystals is the major use of it. The process involves inserting a small crystal into the molten silicon, then pulling it out and rotating it carefully, all under an inert atmosphere such as argon, and in an inert container, such as one made of quartz.

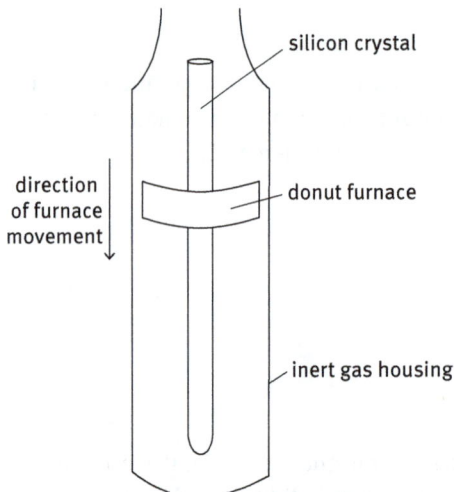

Figure 17.2: Single crystal silicon growth.

A zone-refining furnace can also be used to enhance the purity of silicon. A torus-shaped furnace melts a small section of the silicon crystal, and moves slowly down it. Impurities are concentrated in the melt, and the furnace continues to move downward around the crystal. Since the furnace moves from top to bottom, gravity pulls the impurities downward, and the solid silicon over which the furnace has passed is of higher purity. Repeatedly passing the furnace along the crystal ultimately raises the purity of the silicon and concentrates any impurities in the lowest portion of the crystal. Figure 17.2 shows a simplified diagram of a zone-refining furnace.

17.4 Uses

The reason such large quantities of silicon are refined annually has to do with its use in steel alloys. The USGS states simply: "ferrosilicon accounted for over 60 % of world silicon production on a silicon-content basis in 2022" [1].

A separate document, the USGS Minerals Yearbook Ferroalloys, notes "Domestic consumption of ferroalloys is expected to closely follow the trend in U.S. steel production. Global steel production increased by 5 % to 1.81 billion metric tons in 2018, and demand was expected to increase by 4 % in 2019 and increase slightly in 2020" [11]. Numerous specialty steels can be produced through adjustments in the composition of alloys of ferrosilicon [12]. Thus, it is clear that silicon finds a major use in the steel industry.

Additionally, ferrosilicon finds use in what is called the Pidgeon process, which produces magnesium from dolomite ores. Dolomite is often a mixed magnesium–calcium carbonate. The reaction chemistry can be shown in a straightforward manner, as seen in Scheme 17.3.

$$2\,MgO_{(s)} + Si_{(s)} \longrightarrow SiO_{2(s)} + 2\,Mg_{(g)}$$

Scheme 17.3: Magnesium production using silicon.

What is not shown is that the process is highly energy intensive, and is a high temperature, pyrometallurgical one, which functions by the distillation of gaseous magnesium in a vacuum to produce the solid final product. The starting dolomite must be a fine powder. Calcium oxide in the starting material combines with the silica product (not illustrated in Scheme 17.3) during the reaction and by forming calcium silicate, further drives the reaction to the products side. The reaction is run in stainless steel retorts, and in the cooling area forms what are called magnesium crowns from the pure magnesium vapor. These are then gathered and formed into larger ingots.

The major use of refined magnesium metal is for lightweight alloys or direct use as a metal in automotive applications, where lightweighting of materials is important. US Magnesium, which produces magnesium and magnesium alloys states at its web site: "High purity, corrosion resistant alloys are considered the metal of choice for automotive, power tool, telecommunication and computer component industries" [13].

The average person probably considers the silicon chip in a computer to be the major use for silicon. Concerning this, the USGS further states: "The semiconductor and solar energy industries, which manufacture chips for computers and photovoltaic cells from high-purity silicon, respectively, also consumed silicon metal" [1]. Even though the use is small, high-purity silicon has become vitally important to the developed world.

17.5 Silicones and organo-silicon materials

Carbon is the only element that has been made to polymerize from one atom to another thousands and even tens of thousands of times. Boron does not polymerize, but rather clusters upon itself. Nitrogen does not even oligomerize. Sulfur tends to form eight" membered rings called crowns. Silicon normally does not polymerize from one silicon atom directly to another to anywhere near the number of times that carbon does. But when silicon is polymerized with oxygen atoms, so that the repeat pattern of atoms is silicon-to-oxygen over and over, the resulting materials are a class of polymer that has proven extremely useful in modern society: silicone. More properly named siloxanes, Figure 17.3 illustrates the repeat unit for a silicone.

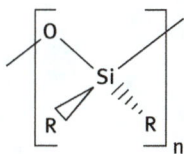

Figure 17.3: Silicone repeating unit.

The R" groups that are the side chains for silicones can be any of a wide variety of organic molecules, and are usually chosen for the macroscopic properties they impart to a finished product. The silicone produced to the greatest extent today is abbreviated PDMS for polydimethylsiloxane, in which the R groups are both methyl groups. This has been characterized as a linear, polymeric material which exists as an oil at room temperature. Thicker, more viscous materials can be produced through the use of R groups that can branch, or that can interconnect. Thus, we see that the repeating backbone of silicones is the inorganic portion of these polymers, but the physical properties that characterize specific silicones are dependent upon the organic side groups.

17.5.1 Silicone production

Scheme 17.4 shows the simplified reaction chemistry that produces a silicone starting material. Note that copper must be present in a catalytic role, although it may be present in as much as 10 % depending on the particular reaction.

$$Si + 2\ CH_3Cl + 2\ Cu \longrightarrow (CH_3)_2SiCl_2 + 2\ Cu$$

Scheme 17.4: Silicone starting material production.

In the past, copper species such as methyl copper and copper-silicon have been isolated, but this isolation of intermediary species is never carried out when the reaction is run on an industrial scale.

The example shown in Scheme 17.4 is for the starting material for PDMS. Other starting materials perhaps obviously require different chlorinated organic molecules.

17.5.2 Silicone uses

Because such a wide variety of chlorinated organics can be converted into silicones, the silicone industry has become one with a wide variety of applications. Producers and manufacturers routinely categorize their products by the particular type of application for which they have been found to be useful. For example, one of the major trade organizations for silicone displays the following categories at its web site [14]:
1. Construction
    –   Protective coatings
    –   Sealants
    –   Other uses

2. Electronics
   - Electronic devices
   - Appliances
   - Power and cabling
   - Others
3. Industrial applications and uses
   - Industrial and mold making
   - Plastics, chemicals, industrial additives
   - Others
4. Life style and personal care
   - Health and medical care
   - Personal and home care
   - Others
5. Transportation
   - Automotive
   - Other (including water-based)
6. Specialty systems
   - Adhesives and coatings
   - Papers and films
   - Textiles and leather
   - Others

Dow Corning, a major producer of silicones, displays a similarly large diversity of applications at its web site [9]. It appears that there are very few classes of materials that have moved so rapidly from laboratory curiosity to major commodity material in so short a time.

17.6 Silanes

Silanes are a class of silicon-containing materials that are similar in formula to the hydrocarbons, but that have silicon atoms in place of the carbon atoms. The general synthesis of the simplest silane can proceed in several different ways. The general chemistry is shown in Scheme 17.5.

$$SiCl_4 + LiAlH_4 \longrightarrow SiH_4$$

Scheme 17.5: Silane synthesis reaction chemistry.

Note that other silicon halides can be used as starting materials, and that calcium aluminum hydride can also be used to effect the hydrogen atom transfer. As well, note the mass balance is not shown in Scheme 17.5. This is because small amounts of impurities are incorporated into the alkali aluminate, usually boron-containing impurities.

But, for an example of another method, Air Liquide states at its web site: "silane is produced from silicon using a two-step process. First, silicon is ground into a powder and then reacted with hydrochloric acid at 300 °C. The result is then boiled with a catalyst (such as aluminum chloride, for example). This reaction produces silane" [15].

We have already seen that the simplest silane (SiH_4) can be used in the production of elemental silicon. It can also be used in the production of solar panels. Other silanes find use as sealants, water repellants, and the protection of surfaces.

17.7 Semiconductors

As the name implies, semi-conductors are materials that conduct electricity, but that do so less well than metals and other conductors. They are not always made with silicon as the main element (germanium is sometimes used), but can be produced using extremely high-purity silicon and often some small amount of another element to adjust the end conductivity. These devices have become so prevalent in modern society that there is now at least one organization devoted to them, their manufacture, and their uses [16].

Semi-conductors are often made to exact specifications by the addition of small amounts of what are called electron-rich or electron-poor additive materials, known as dopants. These dopants generally are categorized as being p-type or n-type, meaning they introduce a space in the material that is lacking an electron, or that has an extra electron, respectively.

Semiconductors are manufactured using the same processes as high-purity silicon, resulting in high-purity cylinders that can be over 100 mm in length. The cylinders are then sliced into wafers. The wafers then must be etched. If a wafer requires doping, the dopant in a vapor phase is passed over the wafer and the dopant deposited on the etched surface, often at temperatures of 650–800 °C. The elevated temperature allows diffusion of the dopant into the silicon wafer. This process requires several hours, as well as an inert gas to maintain the purity of the semiconductor.

17.8 Recycling

The steel recycling industry is a mature one, with scrapyards in most nations dealing in almost all forms of steel. Thus, when steel is recycled, steel alloys made with silicon are also recycled.

High-purity silicon from electronic and computer applications is not routinely re-cycled. Computers contain higher value materials, usually metals, which are recycled in various parts of the world, usually where labor costs are very inexpensive. But the silicon chips in such computers is not recycled or re-used.

Silicones are usually distributed and sold to end users, and the material often ends up in some single, end use. Thus, the recycling of silicones is very difficult. The breakdown of silicones in landfills has proven to be remarkably slow. But there appear to be no reports of discarded silicones causing any health problems in the general population [9].

Bibliography

[1]   United States Geological Survey, Mineral Commodity Summaries, 2023. Website. (Accessed 18 December 2023 as: https://www.usgs.gov, https://doi.org/10.3133/mcs2023, as a downloadable pdf).

[2]   China National Bluestar. Website. (Accessed 23 December 2023, as: http://www.china-bluestar.com).

[3]   Eklem. Website. (Accessed 23 December 2023, as: https://www.elkem.com).

[4]   Ferroglobe PLC. Website. (Accessed 23 December 2023, as: https://www.ferroglobe.com).

[5]   Grupo Ferroatlántica, Spain. Website. (Accessed 23 December 2023, as: https://www.bloomberg.com/profile/company/8933934Z:SM).

[6]   Hankook Silicon Co., Ltd. Website. (Accessed 23 December 2023, as: https://www.enfsolar.com/hankook-silicon).

[7]   JFE Steel. Website. (Accessed 23 December 2023, as: https://www.jfe-steel.co.jp/en/).

[8]   Chevron, acquired Renewable Energy Corp. in February 2022.

[9]   Dow. Website. (Accessed 23 December 2023, as: https://www.dow.com). Acquired Simcala in 20023.

[10] Wacker Chemie AG. Website. (Accessed 23 December 2023, as: https://www.wacker.com/cms/en-us/home/home.html).

[11] USGS 2018 Minerals Yearbook, Ferroalloys. Website. (accessed 23 December 2023, as: https://pubs.usgs.gov/myb/vol1/2018/myb1-2018-ferroalloys.pdf).

[12] Fondelco Group. Website. (Accessed 23 December 2023, as: https://www.fondelcom.in).

[13] US Magnesium LLC. Website. (Accessed 23 December 2023, as: https://usmagnesium.com).

[14] The Society of Silicon Chemistry, Japan. Website. (Accessed 23 December 2023, as: https://www.sscj.jp).

[15] Air Liquide. Website. (Accessed 23 December 2023, as: https://www.airliquide.com).

[16] Semiconductor Industry Association. Website. (Accessed 23 December 2023, as: https://www.semiconductors.org).

18 Lightweight materials

18.1 Introduction

Numerous materials have been created in the past few decades that combine the properties of low density and high strength or durability. Many are made of different metal alloys, while others are based at least in part on nonmetallic elements. There are a number of driving forces for the production of light materials that still have strength and durability. The cost savings for fuel is one such force, especially for the production of airplanes, helicopters, drones, and other flying machines and equipment, in which it is expensive to keep such machines in the air or upper reaches of the atmosphere in terms of maintenance and fuel. Personal protective equipment, clothing, and other materials, when made of lightweight but extremely durable materials, are very beneficial to users. Also, the heavy vehicle manufacturers and military forces of the world continue to look for strong but lightweight materials to be used as the armor on vehicles and other equipment. There exist enough applications for lightweight materials that there is even a trade organization devoted to them and to the dissemination of their many uses [1].

18.2 Lightweight alloys

Alloys that have low density and thus lightweight have been of interest for just over 100 years, since airplanes have evolved from their earliest incarnations, which were largely made from canvas and wood, to those made from different metals, and metal alloys. Aluminum continues to command the largest portion of lightweight metal for the aircraft industry, but several alloys are useful in this industry as well [2]. Indeed, numerous national and international organizations devoted to the applications of aluminum exist [2–9].

18.2.1 Aluminum alloys

A large number of aluminum alloys have been fabricated, mostly since the end of the Second World War, almost always with the idea of creating a strong material but with low density. Additionally, the resistance to corrosion that is displayed by many aluminum alloys makes them useful in applications where the end product will be used in harsh conditions, such as salt water or caustic environments.

Aluminum production was discussed in Chapter 12. As far as alloys, Table 18.1 lists several of the more common aluminum alloys, showing the additive elements to aluminum, as well as major applications of the alloy. The four-number designator for each alloy is the Aluminum Association Wrought Alloy Designation System number. This is a

https://doi.org/10.1515/9783111329512-018

Table 18.1: Aluminum alloys (additive elements in %).

Alloy #	Cu	Cr	Fe	Mg	Mn	Si	Ti	Zn	Zr	Uses
2024	4.3–4.5			1.3–1.5	0.5–0.6					Aircraft
5052	<0.1	0.15–0.35	<0.4	2.2–2.8	<0.1	<0.25		<0.1		Sheet metal, structural
5059	<0.4	<0.3	<0.5	5–6	0.6–1.2	<0.45		0.4–1.5	0.05–0.25	Ships
5456	<0.1	0.05–0.2	<0.4	4.7–5.5	0.5–1.0	<0.25	4.6	<0.25		Automotive, structural
6061	0.15–0.4	0.04–0.35	<0.7	0.8–1.2	<0.15	0.4–0.8	<0.2	<0.25		General purpose
6063	<0.1	<0.1	<0.35	0.45–0.9	<0.1	0.2–0.6	<0.15	<0.1		Window frames, door frames, structures
7050	1.2–1.6	<0.5	<0.5	2.1–2.5	<0.5	<0.5	<0.1	5.6–6.1		Aircraft, marine, automotive
7068	1.6–2.4			2.2–3.0			<0.5	7.3–8.3	0.05–0.15	Military ordinance
7075	1.2–1.6	<0.5	<0.5	2.1–2.5	<0.5	<0.5	<0.5	5.6–6.1		Aircraft

system whereby all aluminum alloys can be classified according to the elements added to produce the material. This table represents only a tiny fraction of the large number of aluminum alloys that have been developed.

18.2.2 Metal foams

Aluminum can also be formed into what are called metal foams, meaning a metal with a significant volume of the end material being air or some other gas in a sponge-like matrix with the metal occupying the solid points of the material, and the gas occupying the empty pockets. While there are metal foams that do not utilize aluminum, this metal is often used because it is already a low-density metal, and the end product – the foam – is designed for some application where low density and mass are important [10].

The production of metal foams involves either injecting a gas into the molten metal or premixing a material with the metal in powder form, then heating the two so that the nonmetal powder component gasifies and is trapped in pockets as the material cools. These techniques, when properly controlled, can produce a metal material that is as much as 75 % gas.

Metal foams have been used as shock absorbent materials, dissipating the force of an impact. This application lends itself to metal foams used in the transportation industry. Additionally, they can be used as heat exchangers. In automobiles and other transportation applications, they also find use in structural applications where weight is a concern. As well, research has been done on the use of metal foams made from bio-compatible materials, such as titanium, in orthopedic applications.

18.2.3 Titanium alloys

Titanium refining and production has been discussed in Chapter 14, but here we will discuss the uses titanium has when incorporated into lightweight alloys. Titanium finds enough uses in lightweight materials that these are tracked by one international trade organization [11]. Most titanium alloys are made for their low density, high strength, and importantly their corrosion resistance. As just mentioned, titanium can be fashioned into a metal foam, but often it is economically more feasible simply to use a titanium-based alloy to provide the proper hard, lightweight, corrosion resistant material for a specific application.

Titanium alloys are as numerous as many other elemental alloys, but can be categorized broadly into alpha alloys, beta alloys, and alpha-beta alloys based on the two allotropic forms of titanium, the alpha and the beta [12]. A selection of the more common titanium alloys are listed in Table 18.2. Numbers for alloying elements are in percentages. The IMI numbers noted in the final column are an older classification system for such alloys, but do not give an indication of the alloying elements. The

Table 18.2: Titanium alloys.

Type	Alloying elements								Comments, characteristics
	Al	Cr	Mo	Nb	Si	Sn	V	Zr	
Alpha									IMI230, 2.5 %Cu, used in aircraft
Alpha									IMI260, 0.2 %Pd, marine environments
Near alpha	8		1				1		
Near alpha	6		0.5		0.2			5	IMI 685, engine components
Near alpha	6		0.5		0.5	3		4	Ti 1100, high strength
Near alpha	5.8		0.5	0.7	0.3	4		3.5	IMI 834, aerospace compressor parts
Near alpha	5.5		0.3	1	0.3	3.5		3	IMI 829, heat resistance, jet engine parts
Alpha-beta	6						4		
Alpha-beta	6		6			2		4	
Alpha-beta	4		4		0.5	4			IMI 551, high strength
Beta	3	3				3	15		
Beta	3		15	3	0.2				Timetal 21 S, cold formable
Beta	3	6	4				8	4	Beta C, high strength

newer system does indicate the alloying elements. For example, it would list IMI 685 as: Ti-6 %Al-5 %Zr-0.5 %Mo-0.2 %Si. While this is more precise, the nomenclature for an alloy becomes longer.

18.2.4 Magnesium alloys

Magnesium, another low-density metal much like aluminum or titanium, has also found numerous uses where the lightweighting of end product materials is important. Magnesium metal production is tracked by the United States Geological Survey, although its alloys are not. Figure 18.1 shows the worldwide production of magnesium in thousands of metric tons, excluding that of the United States, citing proprietary concerns by the one US producer. The same report does indicate though that one company in Utah recovered magnesium from brines from the Great Salt Lake, to a total of approximately 63,500 tons.

The isolation of magnesium from brine involves the use of calcium oxide to form a magnesium hydroxide precipitate, followed by its conversion to magnesium chloride using HCl. The final step, sometimes called the Dow process, is an electrolytic reduction of the magnesium ion to the metal, and concurrent oxidation of the chlorine. Scheme 18.1 shows the simplified reaction chemistry, while Figure 18.2 shows a diagram of the reduction step. Magnesium metal, which forms around the cathode, must be vacuum siphoned out.

The automotive industry is one large industrial sector that uses magnesium in several areas. Concerning this, the International Magnesium Association states in their web

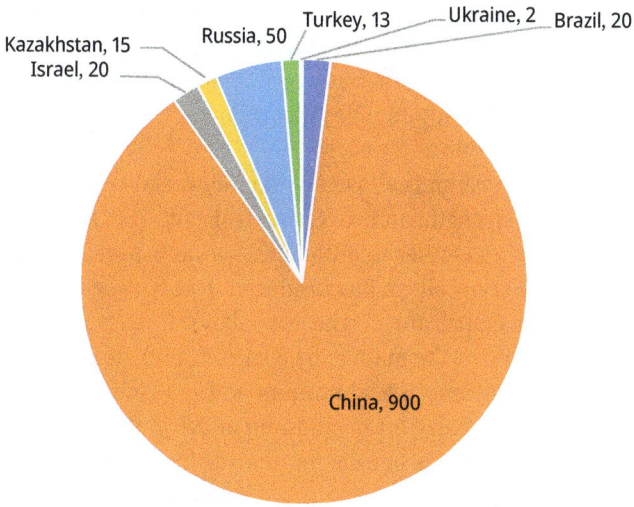

Figure 18.1: Magnesium production, in thousands of metric tons.

$$CaO_{(s)} + H_2O + Mg^{2+}{}_{(aq)} \longrightarrow Mg(OH)_{2(s)} + Ca^{2+}{}_{(aq)}$$

followed by

$$2HCl + Mg(OH)_{2(s)} \longrightarrow MgCl_{2(aq)} + 2H_2O$$

followed by the electrolytic splitting of the salt

$$MgCl_{2(l)} \longrightarrow Mg + Cl_{2(g)}$$

Scheme 18.1: Magnesium production.

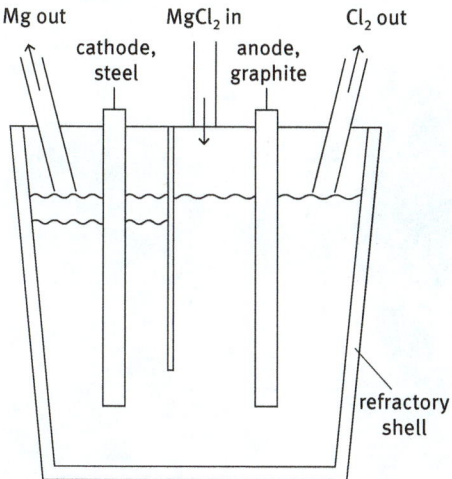

Figure 18.2: Magnesium reduction.

site: "The interest in magnesium use in automotive applications has increased over the past ten years in response to the increasing environmental and legislative influences. Fuel efficiency, increased performance and sustainability are top-of-mind issues" [13]. This sums up succinctly that lighter weight automobiles mean slower consumption of fuels, and thus a better environmental footprint for each auto.

Like other low-density metals, magnesium is alloyed with other metals to produce materials that are low density, high in strength, and corrosion resistant. For these reasons, numerous magnesium alloys find use either in automotive or aerospace applications. Because the development of different alloys has produced a wide range of materials, a systematic code has been developed for magnesium alloys. It is usually two letters and two numbers. The letters indicate the main alloy elements, and the numbers their percentages. For example, AZ61 indicates 6 % aluminum and 1 % zinc. ASTM International has attempted to list and categorize all magnesium alloys under standard B275 [14]. This system is still evolving, as alloys become more exacting, and often are defined to better specificity than a percentage.

18.2.5 Beryllium and beryllium alloys

Beryllium is both a metal with extremely low density and an extremely limited availability worldwide. The United States Geological Survey does track beryllium production, as shown in Figure 18.3, but notes that "Proven and probable bertrandite reserves in Utah total about 19,000 tons of contained beryllium. World beryllium reserves are not available" [15].

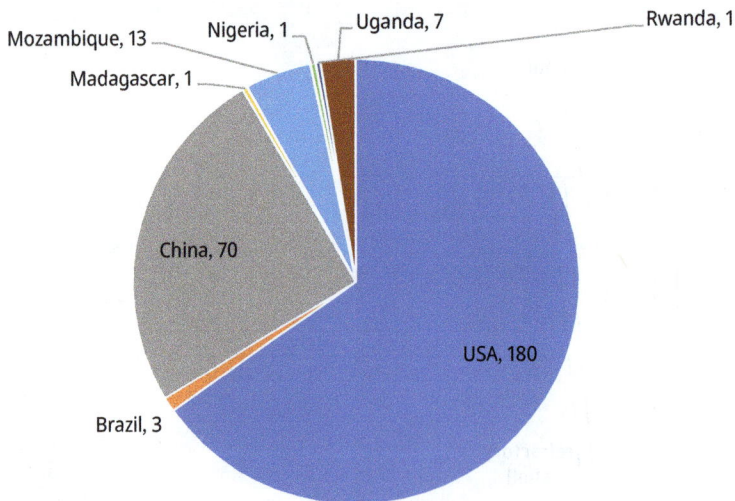

Figure 18.3: Beryllium production, in metric tons.

Note that for beryllium production, the units are in metric tons – and thus become a rather small total when viewed as a global commodity (especially since many other metals are tracked in thousands of metric tons). To get a sense of scale, since the density of beryllium is only $1.85\,g/cm^3$, the 228 metric tons that is the total of Figure 18.2 would occupy a cube only 5.0 m (16.3 ft) on a side, if it were all refined to the reduced metal.

Beryllium is often alloyed with copper, and in various alloys finds several specialized niche uses. The USGS lists the uses of what is called beryllium–copper master alloy as

- "Consumer electronics and telecommunications, 42 %
- Defense-related applications, 11 %
- Industrial components and aerospace industry, 11 %
- Energy applications, 8 %
- appliances, automotive electronics, medical devices and other" 28 % [15]

Perhaps obviously, it is the defense-related applications that are tracked by national governments, since a continued supply of the element becomes important should a nation's military be called into action. What are called beryllium windows are small metal sheets transparent to most X-rays, and thus are extremely useful in various electronic guidance systems. But such X-ray transparency also makes beryllium windows useful for medical and other X-ray diagnostic equipment.

A variety of beryllium alloys exist, ranging from those high in beryllium, such as the e" metals, to those with only a small amount of beryllium, such as beryllium–copper. Table 18.3 provides nonexhaustive list of beryllium alloys. The numbers in columns headed by an element or a compound are percentages.

Table 18.3: Beryllium alloys.

Name	Be	Cu	Co	Al	BeO	Characteristics	Uses
Beryllium copper or beryllium bronze	0.5–3	Remaining				Nonmagnetic	Nonsparking tools
High strength Be-Cu	2.7	Remaining	0.3			High strength	Injection molds
AlBeMet or AM162	*			*		Low density	Aerospace, electronics
E-20	80				20	E Material	Aerospace
E-40	60				40	E Material	Aerospace
E-60	40				60	E Material	Aerospace

*AlBeMet is a proprietary alloy of beryllium and aluminum.

18.3 Aerogels

Aerogels have become a research area of interest after it was found that materials can be made that are as much as 99.98 % air, that have densities which can be as low or lower than air, and that are capable of bearing weight many times greater than their own. The web site aerogel.org divides the broad array of aerogels into subcategories, based on the materials from which they are made, as follows:
-   "Silica
-   Most of the transition metal oxides (e.g., iron oxide)
-   Most of the lanthanide and actinide metal oxides (e.g., praseodymium oxide)
-   Several main group metal oxides (for example, tin oxide)
-   Organic polymers (such as resorcinol-formaldehyde, phenol-formaldehyde, poly-acrylates, polystyrenes, polyurethanes, and epoxies)
-   Biological polymers (such as gelatin, pectin, and agar agar)
-   Semiconductor nanostructures (such as cadmium selenide quantum dots)
-   Carbon
-   Carbon nanotubes, and
-   Metals (such as copper and gold)" [16]

Here we will discuss three of the most common types of aerogels, their syntheses, and applications. The now-traditional means by which they can be produced is referred to as supercritical fluid drying, in which the gel matrix is formed, and the fluid dried around it, producing the target aerogel.

18.3.1 Silica-based aerogels

The production of all types of aerogels is often through a sol-gel process [16]. Because there are numerous types of aerogels, there is a wide variety of different steps that are possible in their production. But the general steps used include the following:
1. Production of a "sol." The sol is a suspension of solid particles in a liquid, with the solid particles being a colloidal dispersion. Ethanol has often been used as the solvent, but other liquids can be used as well.
2. Interlinking through hydrolysis occurs as the sol forms, creating molecular-level bridges of silicon–oxygen–silicon (or metal-oxygen-metal).
3. A catalyst may be added to speed the hydrolysis reactions, the result being the formation of a "gel." The material is defined as a gel when it no longer flows freely as a liquid.
4. Traditionally, the material was subjected to supercritical drying by first exerting pressures and temperatures on it which make it a supercritical fluid, then rapidly decreasing the pressure. This allows the liquid to gasify quickly and evaporate, leaving the network in place.

5. In lieu of rapid pressure decreases, solvent exchange can be used to remove the solvent, again leaving the network in place.

These steps have been adjusted in numerous ways by different researchers and companies, and work for silica-based aerogels as well as those based on other starting materials. BASF for example, has marketed Slentite ®, which is actually a polyurethane-based aerogel [17].

18.3.2 Zirconia-based aerogels

Aerogels based on zirconia can be made in similar fashion to silica-based aerogels, with the obvious difference in starting material. These materials have also found some uses in ceramic blends, because they lighten the weight of the end product or material, and yet still retain significant materials strength.

18.3.3 Titania-based aerogels

A significant amount of research has taken place in the past decade on aerogels based on titania. The robustness of the resultant aerogels, and their ability to be used as catalysts, makes them promising lightweight materials for several applications, including insulators and catalysts. These are a subdivision of what are generally called metal-oxide aerogels, many of which are produced to be used as catalysts.

18.3.4 Aerogel uses

Most aerogels can be used as thermal insulators, simply because their numerous, small open spaces act to insulate against the transfer of heat [17–19]. The largest single use of any aerogel today is that of insulation in the construction of buildings. But aerogels have been tried as insulators in other environments as well, including by the United States Navy as a form of wearable insulation for divers in cold water environments [18].

As well, several types of aerogels are used as catalysts in some role, because of their large surface area.

18.4 Metal-organic frameworks (MOFs)

This new class of materials has seen a significant amount of research in the past 20 years, and holds promise for applications in a number of areas.

While MOFs have yet to capture a large market share, extensive work has been done to study their ability to uptake and store gases such as hydrogen or carbon dioxide, as well as other materials. The ultimate aim is to produce materials that can store hydrogen or carbon dioxide for use as a fuel in hydrogen automobiles, or CO2 from waste streams, and do so without high pressure, or heat-driven systems [20].

18.4.1 Synthesis of MOFs

Because there are a wide number of different MOFs in existence, the synthetic steps in their production will have numerous small variations. But general steps include the following:
1. Hydrothermal mixing of the metal component, which usually becomes a corner moiety, and the organic component, often called the bridging unit, which becomes an edge.
2. When water is not used as a solvent, the term "solvothermal" mixing is used, indicating a nonaqueous solvent.
3. Crystals are allowed to grow slowly from a solution at an elevated temperature.
4. Solvent is evacuated.

After the initial synthesis, MOFs can undergo what is called post-synthetic modification, usually to increase the number of metal ion sites within the open cavity. This increases the binding affinity for small molecules, often gas molecules. This in turn is important because gas storage is a major use for various MOFs.

18.4.2 Uses of MOFs

Uses of MOFs include carbon capture, gas storage, and separations [20,21]. The presence of numerous metal sites within MOF architectures makes them ideal for binding to various polar and nonpolar gases. A recent issue of *Chemical Society Reviews* was devoted to MOFs, and includes methods of synthesis as well as a variety of applications and potential applications [21]. Throughout this, an emphasis is seen for the use of MOFs for gas uptake and release, as well as for catalysis.

18.5 Zeolites

Zeolites have become an important group of lightweight materials that find use in a variety of catalytic roles. Zeolite research is developed enough that there is an international group dedicated to their development and use [22].

Most zeolite syntheses involve hot solution treatment of materials such as silica or alumina with a strong base, such as sodium hydroxide. Such techniques produce high-purity materials suited for a wide range of uses. Indeed, the development of MOF synthetic techniques, mentioned above, developed from the older, more established syntheses of zeolites, and represents an extension of zeolite synthesis from inorganic to a hybrid of inorganic and organic synthesis.

The United States Geological Survey does track natural zeolite production worldwide, through the annual Mineral Commodity Summaries, since these materials find uses in other, important industries. Figure 18.4 shows this in metric tons [15].

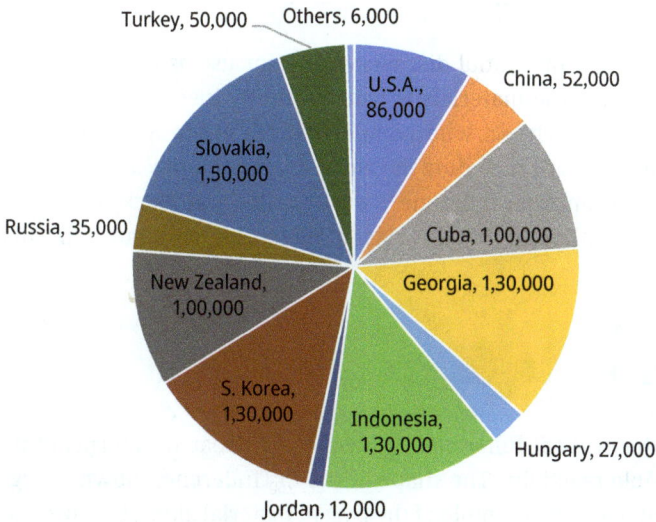

Figure 18.4: Natural zeolite production.

While natural zeolites are not as homogeneous or pure as synthetic zeolites, it can be seen that they are produced on a massive scale, and find a variety of uses. The USGS states: "Domestic uses were, in descending order of estimated quantity, animal feed, odor control, water purification, unspecified end uses (such as ice melt, soil amendment, and synthetic turf), pet litter, fertilizer carrier, wastewater treatment, air filtration and gas absorbent, oil and grease absorbent, fungicide or pesticide carrier, aquaculture, and desiccant" [15]. The first six listed here account for over 70 % of all use.

It may seem curious that natural zeolites are used both as animal feed and as pet litter. As feed, only 0.5 %–2 % of the mass of cattle or chicken feed is the zeolite. This helps keep feed from retaining moisture, and through uptake of nutritionally needed ions, helps produce healthier cattle and chickens. As a pet litter, zeolite material makes up a large percentage of the litter material, and is used to absorb urine and waste, which in turn controls odor, usually in a domestic environment.

Synthetic zeolites exist in more than 200 forms, and find extensive use in the petrochemical industry. The size of zeolite cavities allows specific molecules to bind and others to pass through the material, thus effecting separations. Hydrocracking and catalytic cracking of hydrocarbons are often enabled through the use of zeolites.

As well, synthetic zeolites do find use in the construction industry, much like those mentioned by the USGS Mineral Commodity Summaries, above. They have been used successfully in the production of various types of concrete, where the material must be worked in a semi-liquid state, even in cold temperatures.

18.6 Fullerenes

Few materials have progressed from initial discovery to major use as the fullerenes. The informal nomenclature of these materials is usually intertwined with the name Buckminster Fuller, the famous architect who popularized the geodesic dome structure in several of his building designs including the world-famous Montreal Biosphere. Terms such as "fullerene", "buckminster fullerene", "bucky balls", and "bucky tubes" all come from this connection, and the fact that this class of molecules looks much like his structures.

18.6.1 Fullerene synthesis

First discovered in 1985, the original fullerene, C_{60}, was hotly debated in terms of its characterization and possible reactivity. The shape of the C_{60} fullerene, shown in Figure 18.5, was never determined from a sample of the parent material alone, because the high symmetry of the molecule prevents it from aligning and crystallizing with atoms in each molecule always in the same positions. Rather, an osmium derivative was crystallized, proving the soccer-ball shape of the parent molecule.

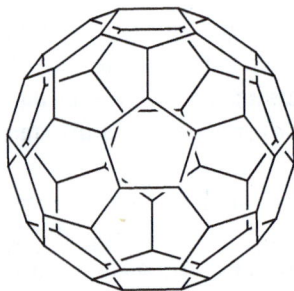

Figure 18.5: Shape of C_{60} fullerene.

The subsequent discovery and development of higher molecular weight fullerenes has quickly given rise to a field of materials that are light in weight (density tends to be $1.7\,g/cm^3$) and robust in a wide variety of uses.

Fullerenes have been synthesized by the electrolysis of a piece of graphite, usually a rod, in a vacuum. The soot is collected and extracted for the C_{60}, and for any other, higher molecular weight fullerenes.

What are called metallofullerenes or endohedral fullerenes are those fullerenes that contain a metal within the fullerene cage. The synthesis of such compounds involves having a small quantity of the element to be incorporated into the cage present in the chamber when electrolysis of the graphite source occurs. The resultant black, sooty materials will be a mix of fullerene and metallofullerenes that will require separation. Nomenclature for this involves using the "@" sign to indicate a metal in the cage. For example, $Hg@C_{60}$ indicates mercury within the C_{60} cage.

18.6.2 Fullerene uses

Fullerenes have found some use in the medical field, in general as contrast agents for various tests. While they have not become a large commodity, there has been enough research in these areas that reviews of the compiled literature have already been written. For example, in the review article "Multifunctional Fullerene- and Metallofullerene-Based Nanobiomaterials", the authors state in their abstract: "Due to their unique physicochemical properties, carbon-based nanomaterials such as fullerenes, metallofullerenes, carbon nanotubes and graphene have been widely investigated as multifunctional materials for applications in tissue engineering, molecular imaging, therapeutics, drug delivery and biosensing" [23]. It appears then that the use of fullerenes and metallofullerenes may in the near future carve a niche into the field of medical diagnostic materials.

Additionally, a company has begun marketing C_{60} in olive oil as a food supplement that has anti-aging properties [24].

18.7 Recycling and re-use

The metal-based materials discussed in this chapter can all be recycled, although the total amount of beryllium used is small enough that such recycling is usually through the company that sold the user-end item. In other words, suppliers ensure recycling or re-use of it. While a significant amount has been written concerning various lightweight materials, their production, and their uses [25], other than high-value metals such as aluminum and magnesium, there appears to be very little that has to do with recycling and second use.

There have not yet been any large-scale attempts to recycle materials such as aerogels, MOFs, zeolites, or fullerenes. Natural zeolites are inexpensive enough that they are generally not recycled.

Bibliography

[1] Lightweight Technology Council – The Green Truck Association. Website. (Accessed 23 December 2023, as: https://www.greentruckassociation.com).

[2] International Aluminum Institute. Website. (Accessed 23 December 2023, as: https://international-aluminum.org).

[3] The Aluminum Association. Website. (Accessed 23 December 2023, as: https://www.aluminum.org).

[4] European Aluminium HOME. Website. (Accessed 23 December 2023, as: https://european-aluminium.eu).

[5] Aluminum International Today. The Journal of Aluminum Production and Processing. Website. (Accessed 23 December 2023, as: https://aluminiumtoday.com).

[6] Aluminum Extruders Council. Website. (Accessed 23 December 2023 as: https://www.aec.org).

[7] Australian Aluminium Council. Website. (Accessed 23 December 2023 as: https://www.aluminium, org.au/).

[8] Japan Aluminum Association. Website. (Accessed 23 December 2023 as: https://www.aluminum.or.jp).

[9] AFSA. Aluminum Federation of South Africa. Website. (Accessed 23 December 2023 as: https://www. afsa.org.za).

[10] Metalfoam.net. Website. (Accessed 23 December 2023, as: http://www.metalfoam.net).

[11] International Titanium Association. Website. (Accessed 23 December 2023, as: https://titanium.org).

[12] RTI International Metals, Inc. Website. (Accessed 23 December 2023, as: www.rtiintl.com).

[13] International Magnesium Association. Website. (Accessed 23 December 2023, as: https://www. intlmag.org).

[14] ASTM International. Website. (Accessed 23 December 2023, as: https://www.astm.org).

[15] United States Geological Survey, Mineral Commodity Summaries, 2023. Website. (Accessed 18 December 2023 as: https://www.usgs.gov, https://doi.org/10.3133/mcs2023, as a downloadable pdf).

[16] Aerogel.org. Website. (Accessed 23 December 2023, as: http://www.aerogel.org).

[17] Aspen Aerogels. Website. (Accessed 22 December 2023, as: https://www.aerogel.com).

[18] United States Navy Experimental Diving Unit Technical Report. NEDU-05-02, 2005. Website. (Accessed 21 May 2015, as: http://archive.rubicon-foundation.org/xmlui/bitstream/handle/123456789/3487/ ADA442746.pdf?sequence=1).

[19] BASF. Website. (Accessed 22 December 2023, as: https://www.basf.com/se/en/who-we-are/ innovation/our-innovations/high-performance-insulation.html).

[20] MOF Technologies. Nuada. Website. (Accessed 23 December 2023, as: http://www.moftechnologies. com).

[21] *Chemical Society Reviews*, Royal Society of Chemistry, 2014. Web archive. (Accessed 23 December 2023, as: http://pubs.rsc.org/en/journals/articlecollectionlanding?sercode=cs&themeid=4e6e7e9f-ed6e-49f8-b3d0-de5949df8056).

[22] International Zeolite Association. Website. (Accessed 23 December 2023, as: http://www.iza-online. org).

[23] Guarav Lalwani and Balaji Sitharaman. Multifunctional Fullerene- and Metallofullerene-Based Nanobiomaterials, *Nano Life*, Vol. 3, No. 3, (2013) 1342003-1 to 1342003-22.

[24] Bucky Labs.com. Website. (Accessed 23 December 2023, as: https://buckylabs.com).

[25] SAE International. Website. (Accessed 23 December 2023, as: https://www.sae.org/publications).

19 Ceramics

19.1 Introduction

The development and use of ceramics – solids that are inorganic, nonmetallic, and produced through heating and cooling of a clay-based material – is one of mankind's oldest specialized industries, with almost all civilizations and cultures producing some kind of ceramic from local clays or soils, usually for pottery, although also for building material. Indeed, ceramic production marks many cultures because of one distinctive style or another, from that of the ancient Chinese, who are believed to have made the world's first pottery roughly 20,000 years ago, to that of the Uruk period of ancient Sumer. Also, highly developed pottery can be found in cultures as widely separated as those of the Minoans and other ancient Greek city-states and the Marajó at the mouth of the Amazon, where enormous ceramic funerary urns have been found [1]. Yet while the output of various clay objects was quite large in some ancient societies, such as the Roman Empire, such production remained a cottage industry until the onset of the Industrial Revolution.

As the Industrial Revolution changed and advanced Europe, the long-standing trade secrets of Chinese porcelain – the ceramic material which is still called "China" today – were unlocked by various European craftsmen, such as Johann Friedrich Böttger of Saxony and Josiah Wedgewood of England. Both men are examples of early entrepreneurs who founded influential companies that transformed how people live and their quality of life [2, 3]. These companies, Meissen Coutre and Wedgewood, respectively, continue to be major players in several areas where ceramic and porcelain materials are used today. Figure 19.1 shows a porcelain medal produced by Meissen Porcelain, honoring Mr. Böttger, the discoverer of what is now called Meissen Porcelain, while Figure 19.2 shows a Wedgewood bowl in their trademark blue jasperware. Although many ceramic

Figure 19.1: Meissen porcelain medal.

https://doi.org/10.1515/9783111329512-019

Figure 19.2: Wedgewood bowl.

and porcelain formulas are proprietary, the red color of the medal, which has become a hallmark of much Meissen-ware, is due to iron compounds in the starting material.

19.2 Production

Almost all companies that produce ceramics use proprietary formulas and starting clays and possibly silica that have been discovered through a rigorous trial-and-error process. There has never been an overarching theory for their production. Nevertheless, there are some ingredients and synthetic steps common to all or most ceramics and some processing steps that are also very common. Ceramic and porcelain manufacturing has become so prevalent an industry that several organizations exist to promote the uses of ceramic materials and exchange information about them [4–11].

19.2.1 Syntheses

Mixing

Clay, sand, and silica portions are mixed, oftentimes with some solvent (not always water), to a consistency that can be worked into a specific shape.

Forming

The ceramic batch is molded and formed into the desired end shape. While some objects are still formed by hand – such as on a pottery wheel – many more are formed with various molds and mechanical apparatus.

Firing

The ceramic object is subjected to high heat, usually in a kiln – a refractory chamber or oven. Kiln temperatures range from 900–1,100 °C in most cases. This firing makes the object hard and robust, and generally inert to other materials, especially liquids, with which it comes in contact.

Some ceramics are still sun-dried, but such materials absorb water when they become wet, while fired ceramics do not. Thus, sun-dried materials find only limited use.

Glazing

After firing, one or more glazes are applied to many ceramic objects. The glaze is often applied as a slurry, with silica as a major, solid component, although some glazes are applied by what is called dry dusting the ceramic surface. Metal oxides can be mixed into glazes to impart various colors. The glaze also further protects the object from unwanted interaction with the environment and degradation. The glaze is often fired onto the ceramic object at or near the same temperature as the initial firing. Figure 19.3 shows examples of ceramics that have been fired.

Figure 19.3: Fired ceramics.

19.2.2 Tailoring properties of ceramics

Since ceramics are produced for applications that require strong materials that are hard, rigid, and able to withstand both pressure and heat, various formulas are produced, then tested to meet such standards. As mentioned, there is as yet no established, theoretical basis for the properties of ceramics. This trial and error method remains the approach used to determine the physical properties of ceramic materials and objects.

Likewise, numerous components are used in glazes, or mixed to produce certain colors. Table 19.1 gives a nonexhaustive list of them.

Table 19.1: Ceramic glazes.

Compound	Color	Comments
Chromium oxide	Green	With other metal oxides, chromium oxide can yield several colors
Cobalt carbonate	Blue	
Cobalt oxide (CoO)	Blue	
Cobalt oxide	Black	When mixed with iron
Copper oxide	Green	
Copper oxide	Copper	When used in lower fire glazes, such as raku
Iron oxide (Fe_2O_3)	Red to brown	
Iron oxide	Cream white	When mixed with tin oxide
Manganese dioxide (MnO_2)	Blue to purple	
Uranium oxide	Red to yellow	
Organic material	Black	Often as ash

19.3 Applications and end uses

As mentioned in the introduction, ceramics have been used for pottery for millennia, and that tends to be what many people consider the major use for all ceramics. But the reality encompasses many more possibilities than pottery containers. The United States Geological Survey does track what it calls the "Domestic Production and Use" of clays, and divides clays into six categories, then listing the uses for each. Table 19.2 provides this breakdown.

Table 19.2: Types and uses of clay.

Type	Chemical formula	Uses [12]
Ball clay	Mixture of kaolinite, mica, quartz	Floor and wall tile 38 %, sanitary ware 20 %, other 42 %
Bentonite	$(Na, Ca)_{0.33}(Mg, Al)_2(Si_4O_{10})(OH)_2 \cdot n\,H_2O$	Drilling mud 30 %, absorbents 27 %, iron ore pelletizing 14 %, foundry sand bond 16 %, other 13 %
Common clay	Variable	Brick 46 %, lightweight aggregate 24 %, cement 20 %, other 10 %
Fire clay	$Al_2O_3 \cdot 2\,SiO_2 \cdot 2\,H_2O$, possibly silica	Heavy clay products 57 %, refractory products and other 43 %
Fuller's earth	$(Al, Mg)_2Si_4O_{10}(OH) \cdot 4\,H_2O$	Absorbents 75 %, other 25 %
Kaolin	$Al_2Si_2O_5(OH)_4$	Paper 50 %, other 50 %
Kyanite	Variable	Refractories 90 %, other 10 %

It is clear from this that several types of clay are not used for ceramics or porcelain at all. Yet all these materials find large-scale use in different industries. Examples of large-scale industrial use, or the mass production of user end items, all of which involve some ceramic formulation, include the following:
- Common bricks
- Sewer and drain pipes
- Drain and roofing tile
- Flower pots
- Refractory lining

19.4 Recycling and re-use

The recycling of ceramic materials is seldom undertaken, simply because most of these materials have been fired, and thus made into what is considered a permanent material or end use item. For materials that are used on a large scale, there are instances in which the broken, crushed material can find a second use, perhaps as a filler material in roadways, highway berms, or base material in construction projects.

Bibliography

[1]   Margaret Young-Sánchez and Denise P. Schaan. Marajó: Ancient Ceramics from the Mouth of the Amazon. 2011, Denver Art Museum. ISBN: 978-0-914738-73-2.
[2]   Meissen Coutre. Website. (Accessed 24 December 2023, as: https://www.meissen.com/net).
[3]   Wedgwood. Website. (Accessed 24 December 2023, as: https://www.wedgwood.com).
[4]   The Tile Council of North America. Website. (Accessed 24 December 2023, as: https://tcnatile.com).
[5]   TCNA: Building Green? Website. (Accessed 24 December 2023, as: https://tcnatile.com/wp-content/uploads/2022/08/Building-Green-with-Tile.pdf).
[6]   United States Advanced Ceramics Association. Website. (Accessed 24 December 2023, as: https://advancedceramics.org).
[7]   Australian Clay Minerals Society, ACMS. Website. (Accessed 24 December 2023, as: http://www.smectech.com.au).
[8]   Clay Science Society of Japan. Website. (Accessed 24 December 2023, as: http://www.cssj2.org).
[9]   European Clay Groups Association (ECGA). Website. (Accessed 24 December 2023, as: twitter.com/EU_ClayGroup).
[10] International Association for the Study of Clays (AIPEA). Website. (Accessed 24 December 2023, as: https://aipea.org).
[11] The Clay Minerals Society. Website. (Accessed 24 December 2023, as: https://www.clays.org).
[12] United States Geological Survey, Mineral Commodity Summaries, 2023. Website. (Accessed 18 December 2023 as: https://www.usgs.gov, https://doi.org/10.3133/mcs2023, as a downloadable pdf).

20 Hard materials

20.1 Introduction

Hard materials are a field of inorganic chemistry or materials chemistry that has seen rapid expansion throughout the 20th century, but particularly rapid growth since the Second World War. Perhaps the two most obvious cases where hard materials are required are in the development of better, high-performance engines and the development of modern armor. The first case involves the higher strength and durability materials that have become required as engines were developed for high-speed automobiles, trains, ships, and planes – the lattermost having gone from a simple engine that makes a vehicle fly to the modern jet engines used in both civilian and military aircraft in a matter of only four decades. The second case has developed as what is called the "lethality of warfare" has increased through two world wars and several small wars which all fell under the larger category of the Cold War. As better, more penetrative bullets and projectiles were produced, some form of modern armor had to be created to stop or deflect them.

This broad field known as hard materials can be broken down into several sections. Hard metal alloys have already been covered in preceding chapters. Ceramics were discussed in Chapter 19, but some of the materials discussed here are ceramic in nature. Diamonds, carbides, borides, and nitrides all are encompassed in this broad category.

20.2 Diamond

20.2.1 Introduction

There are several different natural materials and objects that have invoked wonder throughout history. Diamonds are certainly one such. Alluvial diamonds have been gathered from rivers for millennia, and such work is still undertaken today [1]. Many natural diamonds however are extracted from kimberlite deposits. Large, gem-grade diamonds are often given names, such as the Koh-i-Noor, the Great Star of Africa, or the supposedly cursed Hope diamond, because they are considered so special and rare. But natural diamonds are each unique, and often contain some small flaw or series of small flaws in their crystal lattice that makes them unsuitable for use in an industrial setting. As well, they are rare enough that historically it has not been profitable to use what are called gem-grade stones in most industrial environments. Their value however, is such that a trade organization exists promoting the various uses of diamonds [2].

Diamonds do have several important industrial uses that we will examine, as well as their well-known use in jewelry, and we will therefor discuss artificial, or man-made, diamonds [3, 4].

https://doi.org/10.1515/9783111329512-020

20.2.2 Diamond synthesis

The production of synthetic diamonds is both a mature industry and one that is still evolving in terms of the size of the diamonds that can be produced, and the purity or level of inclusion of directed impurities. Small diamonds with specific uses in industry have been known for decades, and while methods of production are constantly being examined, the established synthetic method remains one of high temperature and high pressure.

The USGS Mineral Commodity Summaries track the production of diamond annually, and note that approximately 99 % of industrial use of diamonds is in the form of synthetic diamonds in what is called "bort, grit, and dust" [5]. Further, it states: "In 2014, total domestic production of industrial diamond was estimated to be 108 million carats with a value of $ 73.2 million" [4]. Since diamond is the only commodity measured in carats, a sense of scale is easier to grasp if this is converted into a metric unit, such as kilograms. One carat equals 0.0002 kg, and thus 108,000,000 carats equals 21,600 kg (or 47,520 lb). While this is a very small overall total when compared to the metals discussed in Chapters 11 to 17, it is still a significant commodity when the diverse uses of industrial diamond are considered. Uses of industrial diamonds include the following:
–   Cutting edges of cement saws
–   Computer chip production
–   Machinery manufacturing
–   Mining drills
–   Stone polishing [5]

20.3 Silicon carbide

Silicon carbide has a relatively short history. Also known as carborundum, as well as moissanite, it is a very rare naturally occurring mineral. All silicon carbide that is used industrially today is synthetic, and is usually produced by the high temperature combination of carbon and silica (at approximately 1,700–2,500 °C) in an electric furnace with a graphite core. This is still called the Lely process or the Acheson process, and can produce SiC crystals of greater than 1 cm.

Silicon carbide finds wide use as an industrial abrasive, and is often considered a substitute for diamond grits in this regard. The United States Department of Defense bothers to track the use of silicon carbide because of its strategic value [6], and lists it among 76 different materials that are tracked, "because of their important defense uses and possible fragility of supply" [6].

While there are numerous producers of silicon carbide, and their applications profiles do differ, one of those firms, Washington Mills, lists its applications as: "advanced ceramics, anti-skid, blasting media, bonded abrasives, brake pad, buffing compounds, coated abrasives, coatings, cosmetics, electronics, glass, heavy media separation, investment casting, laminates, lapping, metallurgical, mass finishing, pressure blasting, re-fractory, solar, surface preparation, thermal spray, wire sawing." Clearly, silicon carbide is used as much more than grit and industrial abrasives [4, 5].

20.4 Boron and carbon nitrides

Boron nitride, another exceptionally hard material, is used in a variety of different industrial applications, many of them involving either materials that must be able to endure high temperatures, high stress, or both.

While there are different forms of boron nitride, one of the most common is hexagonal boron nitride, or h-BN. This is actually the softer form of the material, and finds use as a lubricant. Its synthesis can be shown according to the following reactions, in Scheme 20.1.

$$2\,NH_3 + B_2O_3 \longrightarrow 2\,BN + 3\,H_2O \qquad\qquad \text{at } 900\,°C$$
also
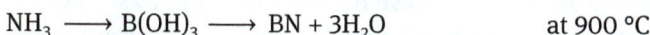
$$NH_3 \longrightarrow B(OH)_3 \longrightarrow BN + 3H_2O \qquad\qquad \text{at } 900\,°C$$
also

$$B_2O_3 + CO(NH_2)_2 \longrightarrow CO_2 + 2\,BN + H_2O \qquad\qquad \text{at } 1{,}000\,°C$$
also

$$B_2O_3 + 10\,N_2 + 3\,CaB_6 \longrightarrow 20\,BN + 3\,CaO \qquad\qquad \text{at } 1{,}500\,°C$$

Scheme 20.1: Synthesis of boron nitride.

These reactions are run under nitrogen to ensure a minimal amount of co-manufactured side products. The major by-product is unreacted boron oxide which can be evaporated at temperatures higher than the reaction conditions. This results in boron nitride purities at or near 99 %. If end use items need to be made from boron nitride, a high-temperature high-pressure system is used to form them. As mentioned, h-BN can be used as a lubricant. Other uses include:

- Abrasives
- Alloy component
- Personal care products
- A component in paints
- Ceramic additive
- Synthetic rubbers
- Plastics

Cubic boron nitride (c-BN) is a hard material that is structurally close to diamond. Designated c-BN, it is often produced by applying high heat and pressure to existing h-BN (pressures of approximately 10 GPa and temperatures of 2,000–3,000 °C). One major use of c-BN is drills for steel, because the material is not soluble in steel, while diamond – being made of carbon – is soluble in steel when both the drill and the steel being drilled are at high-enough temperatures.

Enough boron nitride is used each year that there are trade names for it. Borazon is one for c-BN, the term being a trademark name at General Electric, where it was first manufactured [7].

Carbon nitride has also been theorized in the past as a hard material, and what is called beta carbon nitride – β-C_3N_4 – has been prepared in the past twenty years. This material has not yet been made in large enough quantities to have established any niche industrial use however.

20.5 Metal borides

There exist other metal borides as well. They have been of interest at least since the 1950s, when patents were first filed for them [8]. Initially, metal borides were manufactured by the high-temperature reaction of a metal oxide, boron carbide, and carbon. Zirconium boride (ZrB_2), osmium boride (OsB_2), rhenium boride (ReB_2) and ruthenium boride (RuB_2) have all been manufactured, and have found uses related to their hardness. As well, some metal borides, such as ReB_2 exhibit metal-like conductivity, and may find applications related to this property.

20.6 Recycling

In general, the materials discussed here are not made in large enough quantities that their recycling has proved to be economically feasible, with the exception of synthetic diamonds.

Concerning diamond recycling, the USGS Mineral Commodity Summaries states: "In 2014, the amount of diamond bort, grit, and dust and powder recycled was estimated to be 44.1 million carats. Lower prices of newly produced industrial diamond appear to be reducing the number and scale of diamond stone recycling operations. In 2014, it was estimated that 390,000 carats of diamond stone was recycled" [5]. As with many materials that are recycled, the driving force appears to be an economic one.

Bibliography

[1] Alluvial Diamond Mining Project. US Geological Survey. Website. (Accessed 24 December 2023, as: https://www.pubs.usgs.gov/sir/2010/5044).

[2] DeBeers Group. Website. (Accessed 24 December 2023, as: https://www.debeersgroup.com).

[3] World Diamond Council. Website. (Accessed 24 December 2023, as: https://www.worlddiamondcouncil.org).

[4] United States Geological Survey, Mineral Commodity Summaries, 2023. Website. (Accessed 18 December 2023 as: https://www.usgs.gov, https://doi.org/10.3133/mcs2023, as a downloadable pdf).

[5] Washington Mills. Website. (Accessed 24 December 2023, as: https://www.washingtonmills.com/silicon-carbide).

[6] United States Department of Defense. Strategic and Critical Materials 2013 Report on Stockpile Requirements, 2013. (Accessed 24 December 2023. Downloadable at: mineralsmakelife.org/assets/images/content/resources/Strategic_and_Critical_Materials_2013_Report_on_Stockpile_requirements.pdf).

[7] General Electric. Website. (Accessed 24 December 2023, as: https://www.ge.com).

[8] Method of Making Metal Borides. United States Patent: US 2957754A.

Index

https://doi.org/10.1515/9783111329512-021